TECHNICAL REPORT

Allocation of Forces, Fires, and Effects Using Genetic Algorithms

Christopher G. Pernin, Katherine Comanor,
Lance Menthe, Louis R. Moore, Tim Andersen

Prepared for the United States Army

RAND ARROYO CENTER

The research described in this report was sponsored by the United States Army under Contract No. W74V8H-06-C-0001.

ISBN: 978-0-8330-4479-2

The RAND Corporation is a nonprofit research organization providing objective analysis and effective solutions that address the challenges facing the public and private sectors around the world. RAND's publications do not necessarily reflect the opinions of its research clients and sponsors.

RAND® is a registered trademark.

Published 2008 by the RAND Corporation
1776 Main Street, P.O. Box 2138, Santa Monica, CA 90407-2138
1200 South Hayes Street, Arlington, VA 22202-5050
4570 Fifth Avenue, Suite 600, Pittsburgh, PA 15213-2665
RAND URL: http://www.rand.org/
To order RAND documents or to obtain additional information, contact
Distribution Services: Telephone: (310) 451-7002;
Fax: (310) 451-6915; Email: order@rand.org

Preface

This report is part of a project titled "Representing the Allocation of Forces, Fires, and Effects Using Genetic Algorithms." The project strives to develop appropriate representations of intelligence-driven command and control for use in U.S. Army constructive simulations. This report, "Allocation of Forces, Fires, and Effects Using Genetic Algorithms," explores a method for representing sophisticated command and control planning algorithms that specifically concern maneuver planning and allocation schemes. The report describes a model developed within the RAND Corporation that uses genetic algorithms to compute the allocation of forces, fires, and effects with detailed look-ahead representations of enemy conduct. The findings should be of interest to those concerned with the analysis of command, control, communications, computers, intelligence, surveillance, and reconnaissance (C4ISR) issues and their representations in combat simulations.

This research was supported through the Army Model and Simulation Office (AMSO)–formed C4ISR–Focused Area Collaborative Team (FACT). It was sponsored by the U.S. Army DCS/G-2 and conducted within RAND Arroyo Center's Force Development and Technology Program. RAND Arroyo Center, part of the RAND Corporation, is a federally funded research and development center sponsored by the United States Army.

The Project Unique Identification Code (PUIC) for the project that produced this document is DAMII05006.

For more information on RAND Arroyo Center, contact the Director of Operations (telephone 310-393-0411, extension 6419; fax 310-451-6952; email Marcy_Agmon@rand.org), or visit Arroyo's Web site at http://www.rand.org/ard/.

Contents

Figures

Tables

Summary

Decisionmaking within the Future Battle Command structure will demand an increasing ability to comprehend and structure information on the battlefield. As the military evolves into a networked force, strain is placed on headquarters and others to collect and utilize information from across the battlefield in a timely and efficient manner. Decision aids and tools on the battlefield, as well as solution methodologies in constructive simulations, must be modified to better show how this information affects decisions.

This report demonstrates how a particular algorithm can be adapted and used to make command and control decisions in analytical planning tools and constructive simulations. We describe a model that uses genetic algorithms in the generation of avenues of approach (AoAs), also called "routes" and "paths," and the allocation of forces across those AoAs.[1]

The model is a representation of planning because it uses intelligence products to determine a preferred route or set of routes and allocate forces to those routes. The intelligence products consist of (1) information about Blue forces, such as their mission and location; (2) information about Red forces, such as their location, capability, intent, and activity; and (3) information about the environment.

We used a genetic algorithm to stochastically search the vast space of possible maneuver schemes (routes) and allocations of forces to these routes. This technique is appropriate for such decisionmaking not only because it can quickly search a very large space but also because it can find "good," feasible, although suboptimal, solutions without becoming mired in an optimization routine. Because real-world problems are not, generally speaking, convex, we are not guaranteed to find global optimal solutions. Hence, some sort of heuristic, such as a genetic algorithm, is required. To improve the efficiency of our search, we broke the model down into two phases, each of which has a genetic algorithm at its core. The first phase discovers potential routes; the second determines a desirable allocation of forces to those routes.

The model developed in this report is unique in many respects. It incorporates many higher-level intelligence products, such as intelligence about location, activity, intent, and capability, into the planning algorithm.[2] It also includes information about the intelligence capability and adaptability of the adversary. Although many of these parameters are largely conceptual, our descriptions of them are not. We quantify each parameter in the model and hence parameterize the battlefield.

---

[1] We use the terms *AoA*, *route*, and *path* synonymously throughout this report.

[2] For additional information about these information products and the fusion of such information, see Pernin et al. (2007).

The employment of such a diverse set of intelligence products allows for sophisticated look-ahead representations of Red forces as opposed to the "static" snapshot representations that are typically used in planning sessions. These more-sophisticated representations of Red forces allowed a diverse set of outputs to occur (see Figure S.1), including the discovery of Blue tactical "feints" with no hardwired heuristic to drive the solution to this famous military tactic.

The model also features terrain representations that affect each side's ability to move and hide. Terrain is characterized by three parameters: impassibility, inhospitableness, and shadowing. "Impassibility" measures the degree to which Blue finds certain terrain difficult to cross. "Inhospitableness" measures the degree to which Blue believes that terrain will affect Red's choice of location. "Shadowing" models the influence of terrain on Red's effect on Blue. As an effect on effect, shadowing is a second-order consideration, and although we have included a description of the effect of shadowing in this report, we do not currently implement the shadowing function in our computer model.

To validate the model, we first considered a simple scenario in which Red was stationary and located at the midpoint between Blue's start and end points. Blue's mission was to reach his destination while minimizing his exposure to Red. Routes discovered by the model show that the chosen path depends on intelligence about Red's location and capability. Specifically, if Blue is relatively certain about Red's location and capability, then Blue benefits from maneuvering around Red. However, if Blue is very uncertain about Red's location or capability or both, Blue does not benefit from maneuvering around the enemy, but rather should take the most direct route to his destination. Over various cases, the model clearly demonstrates the value of intelligence in the planning process.

Figure S.1
Tactical Feint as Blue Plans to Move from Point A to Point B to Avoid a Mobile Red

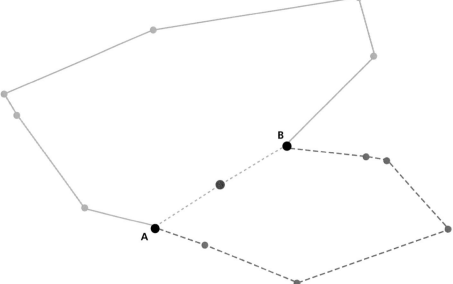

NOTE: Three separate paths are shown in this example. Red starts at the position shown between points A and B.

The model also demonstrates the influence of terrain on AoA selection. We modeled a simple mountain feature that affected both Blue and Red forces. Because Red was less likely to travel over the mountainous area, his effect on Blue was diminished in this region. Hence, we expected this region to be more desirable to Blue. However, the desirability of Blue routes was also penalized by the difficult terrain. Ultimately, the model discovered Blue AoAs that avoided the mountainous region as much as possible while minimizing Blue exposure to Red.

We also considered various other cases to demonstrate the effect of enemy intelligence and adaptability on AoA selection. Enemy adaptability is the rate at which Red receives updates about Blue's route. Enemy intelligence is the amount of information Red receives at each update. Unsurprisingly, Blue is more likely to evade a less adaptive Red than a more adaptive Red. Also, more-intelligent Red forces diminished Blue's options.

We also demonstrated the effect of Blue knowledge of Red activity on Blue force allocation to AoAs. Activity knowledge is the likelihood that Blue knows to which AoA each Red unit has been assigned. We found that with partial activity knowledge, Blue tries to dominate Red where Blue expects Red to be; Blue will also allocate forces where he expects Red not to be. With perfect activity knowledge, Blue can completely avoid the AoA where the dominant Red force is located.

Our model should be of use to those considering command and control representations in combat simulations.

Abbreviations

AMSO   Army Model and Simulation Office

AoA   avenue of approach

C2   command and control

C4ISR   command, control, communications, computers, intelligence, surveillance, and reconnaissance

FACT   Focused Area Collaborative Team

GA   genetic algorithm

ISR   intelligence, surveillance, and reconnaissance

MDMP   Military Decision Making Process

NN   neural network

SA   situational awareness

SOM   self-organizing map

CHAPTER ONE

Introduction

Decisionmaking within the Future Battle Command structure will demand an increasing ability to comprehend and structure information on the battlefield. As the military evolves into a networked force, strain is placed on headquarters and others to collect and utilize information from across the battlefield in a timely and efficient manner. Decision aids and tools on the battlefield, as well as solution methodologies in constructive simulations, must be modified to better show how this information affects decisions.

The problem with the progression to an information-centric force lies in how best to incorporate all relevant pieces of information about both Blue and Red while making command decisions. This report demonstrates one method for incorporating sophisticated intelligence information on Red location, capabilities, and intent into the generation of Blue plans. Specifically, we describe a model that uses genetic algorithms to determine avenues of approach (AoAs) and the allocation of forces and effects across those AoAs.[1]

The generation of AoA and force-allocation schemes fits generally into the command and control (C2) research that is part of a larger portfolio of projects at RAND (see Pernin et al. [2005] and Pernin et al. [2007], for examples). The overarching goal of this body of research is to forge a strong analytical linkage between C2, communications, computers, and ISR (C4ISR) and operational outcomes (see Figure 1.1). This linkage is meant to work in concert with other processes for use in constructive combat simulations and other analytical devices. This report makes use of intelligence products generated in the fusion process (Pernin et al. [2007]) in new C2 decisionmaking algorithms for the allocation of forces, fires, and effects.

Figure 1.1
The Linkage Between Intelligence, Surveillance, and Reconnaissance (ISR) and Operational Outcomes

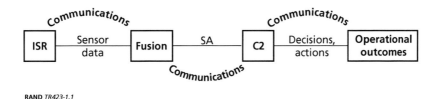

RAND *TR423-1.1*

---

[1] We use the terms *AoA*, *route*, and *path* synonymously in this report to refer to geographical lines of progression that a unit, group, or person might take to get from one point to the next. The term *scenario* captures the larger context in which the path is being taken. Thus, in a given scenario (such as a full frontal assault on enemy positions), multiple paths might be followed by different units advancing in the assault.

Approaches to both the modeling and execution of C2 are evolving. In the past, many constructive simulations used scripted or rule-based approaches in which subject matter experts determine, to some degree of precision, the results of many decisions before the simulation is run. These rules do not change during the course of the simulation, and although they may be appropriate early in a model run, they do not necessarily remain so as the simulation evolves.

More-recent models of C2 use valuation approaches in which information generated during a run is used to calculate parameters for decisionmaking. Myopic strategies base their decisions on the model state at the current time step in a combat simulation. These types of valuation strategies can be easy to implement, but their ability to provide satisfactory results many time steps later is often limited. Look-ahead strategies, on the other hand, estimate future states and make decisions based not only on the current state but also on expectations of future states. Forecasting these expected future states is challenging. Given the large space of possible outcomes, an efficient means of searching this space is required. In an example described by the Military Decision Making Process (MDMP), three alternative courses of action are generated from analysts who are able to reasonably generate future enemy tactics and maneuver.

Algorithms Considered

Allocating forces, fires, and effects to AoAs creates a very large search space in which an analytical tool must represent decisionmaking. Hence, we began our research by exploring a variety of potential algorithms, including various heuristics, neural and Bayesian networks, fuzzy logic, and greedy and genetic algorithms. This report details our focus on the development of a genetic algorithm (GA) as a means of efficient stochastic search. Table 1.1 briefly summarizes the algorithms that we considered and presents some justification of our decision to use the GA.

Table 1.1
Many Decision Algorithms Are Available

Type	Description	Useful Applications
Greedy algorithm	Makes decisions sequentially based on what seems best at the moment	Deciding a route via a series of individually low-cost moves
Artificial neural network	Mimics neural circuitry; learns by example	Deciding enemy intent based on formation
Bayesian belief networks	Makes inferences about the likelihood of events using Bayesian statistics	Deciding enemy objectives given an enemy state
Fuzzy logic	Determines partial-set memberships of system states	Deciding whether units are correctly spaced
Genetic algorithm	Evolves a population of solutions via natural selection	Deciding a route based on overall cost

Neural Networks

Neural networks (NNs) come in many varieties, but each rests on the concept of a neuron as a unit for information storage and mapping input to output. A single neuron usually receives a numerical input vector that is either binary or part of a continuum. Each element of the input vector is scaled by a weighting constant, which essentially assigns a degree of importance to each input. The result of the dot product is then entered into a "squashing" function whose output, a number between either 0 and 1 or –1 and 1, is then sometimes used as the input to another neuron.[2] NNs are just one of a number of connectionist approaches to modeling.

Neurons are connected together in many different ways to develop NNs. The most common connection type is the 3-layer, feed-forward network, in which a row of input neurons, each of which takes one input with one weight, passes its output to another layer called the "hidden" layer (because its output is not shown). The hidden layer computes and then passes its outputs to the output layer, which performs a final computation before giving the answer. Feed-forward networks (sometimes recurrent) are useful for classifying inputs into a small number of slots. Other types of networks are self-organizing maps (SOMs) in which neurons are connected together in a grid such that each neuron is connected only to its neighbors, receiving input from the bottom and giving output at the top. SOM-like networks excel at picking out features from images. Other network types include Hopfield networks (recurrent) and Boltzmann machines (stochastic).

NNs can be trained to produce specific outputs for specific inputs and also to produce specific answers for specific kinds of inputs. This leads to their most common usage: pattern recognition. Their status as a decision algorithm rests on their ability to classify inputs for which they have not been previously trained. The greatest disadvantage of NNs is that they are exceedingly slow to train because they are usually run on a single processor computer and do not take advantage of their massive parallel processing potential—the potential that nature maximizes in human brains. We see the same problem later in GAs.

NNs are out of favor as a decisionmaking algorithm because they lack computational efficiency and tend to act as a "black box" unless a laborious query-and-response procedure is undertaken to develop rules after training is complete. (After training, the network can be discarded.)

NNs have been successfully applied to automatic target recognition (Rogers et al. [1995]) and data fusion (Bass [2000]; Filippidis et al. [2000]) and in agent-based, recognition-primed decision models (Liang et al. [2001]). Other applications include synthetic aperture radar image classification (Qin et al. [2004]) and determining decisive points in battle planning (Moriarty [2000]).

Bayesian Belief Networks

Bayesian belief networks are another example of a connectionist approach to decisionmaking.[3] In this case, the network is designed in accordance with expert knowledge instead of trained. Belief networks allow users to develop a level of confidence that a particular object will be in a particular state based on certain available information. For example, suppose a person in a house comes to a door whose handle is hot and sees smoke coming from under the door. That

---

[2] Sometimes a Heaviside function is used instead. When continuity or differentiability are required, a sinusoid or an inverse hyperbolic tangent is chosen.

[3] See Krieg (2001) for a short, referenced tutorial on Bayesian belief networks and their application to target recognition.

person might reasonably infer that there is a fire on the other side. Two pieces of information—the hot door handle and the smoke under the door—support the inference. Belief networks, in fact, take this idea a step further and add probability to facts and inferences. The particular weight attached to each fact indicates how much credence the fact lends to an inference. So, a hot door handle may indicate a 50 percent chance of fire, while smoke indicates a 75 percent chance.

Bayesian networks are "directed acyclic graphs over which is defined a probability distribution" (Starr et al. [2004]). Each node in the graph represents a variable that can exist in one of several states: For instance, a node could be "ground forces" and its states could be "attacking," "withdrawing," or "defending." The network is set up to represent causal relationships. If node 1 causes node 2, then we say that node 1 is the parent of node 2. For instance, an "enemy intention" node might be the parent of a "ground forces" node. Bayesian networks can be solved in several ways using conditional probability methods. If we know the "ground forces" state is "attacking," this may give us an inference about the "enemy intention." Conversely, we can use "enemy intention" to try to predict whether the ground forces will attack. Either way, we are using what we know to infer what we do not know.

Bayesian networks work best in domains where variables have a small number of states. They could be useful in multiresolution models where smaller Bayesian networks can be connected into larger Bayesian networks and treated as "black boxes." They are not a good choice for maneuver or force allocation because of their scalability limitations. Probabilities must be defined for each node's input state or own-state pair, and these must be assigned by hand.

Fuzzy Logic

Fuzzy logic aims to represent logic in a more "human" way, which is to say that situations are not always decided 100 percent one way or another. Fuzzy logic is perhaps the simplest decision method and is quite useful when combined with other methods. Invented by Lotfi Zadeh in the 1960s at the University of California, Berkeley, this method defines partial-set memberships on system states.[4] For instance, in their paper on a fuzzy-genetic design for Blue-unit spacing given an attacking Red force, Kewley and Embrechts (1998) mapped unit distance into two fuzzy sets, a "too close together" set and a "too far apart" set, each defined by the set of distances from 0 to an infinite number of meters. The distance between two Blue units was calculated and then assigned a level of membership in each set. Any distance over 10,000 m was too far apart, having a probability of 1, but below that distance the set membership level decreased linearly until 5,000 m, at which point membership dropped to 0. Between 5,000 and 500 m, units were considered correctly spaced and had membership in neither set. At 500 m, the probability of membership in the "too close together" set increased linearly until 250 m, when it reached 100 percent. There is a continuum of set membership into which humans classify objects and events.

Fuzzy logic, although powerful when combined with other methods, requires a lot of manual trial and error, and the risk of designer bias in implementation is greater compared to trained methods such as GAs and NNs. On the other hand, if a problem is defined in a way that makes the set membership function obvious, then it becomes a much better choice and simpler to implement.

---

[4] For example, see Bellman et al. (1964).

In maneuver planning and force allocation, fuzzy logic's usefulness comes from its ability to synthesize easy-to-understand statements from complex data, a kind of fusion. Instead of saying that units are 7,000 m apart, which is a fact, fuzzy logic gives an opinion that is more useful: The forces are 50 percent too far apart. This leads to the judgment that they ought to be closer together. Thus, fuzzy logic allows facts to be translated into judgments quite easily. What fuzzy logic is not good at is telling a unit to go to a particular point (e.g., specific coordinates). It can tell a unit to "go left" or "turn around," given some input data, but these are local judgments based on current circumstances. Global judgments are not possible using fuzzy logic unless another algorithm, such as a GA, is included.

A* and Other Greedy Algorithms

The gaming community relies heavily on greedy algorithms for determining paths through cost topologies. The algorithm known as A* (pronounced "ay-star") is the most common. It combines the Best-First-Search algorithm, a quick algorithm, and Dijkstra's route-finding algorithm, which is an optimal solution-finding algorithm. A* works by starting at a node and using a heuristic to determine the best node to move to from its present node. The choice of this heuristic can depend on many things. For certain choices, A* will behave exactly like Dijkstra's algorithm, testing every path. The downside of A* is that, while it is an excellent route finder, it requires that the designer choose a heuristic, which can lead to rather suboptimal paths if those paths are chosen poorly (Hart et al. [1968]).

Genetic Algorithms

GAs are useful in many applications.[5] However, their reliance on initially random populations prompts one question: How are they different from random sampling? If we could derive a good solution to a problem with a large initial population of samples, we would not need a GA. We could save time by implementing mating and mutation and simply generate and evaluate thousands of potential solutions. Some problems are indeed amenable to random problem solvers of this kind and do not require GAs. However, when the fitness landscape contains high, narrow peaks and wide stretches of barren waste between them, GAs are necessary. If the area covered by fitness peaks approaches zero compared to the number of bad solutions in the landscape—i.e., if good solutions are exceedingly rare—a random problem solver will rarely find a good solution. Such is the case in the natural world, where only a handful of configurations of molecules are reproductive out of the vast numbers of other configurations. These fitness landscapes correspond to the "difficult" problems where traditional algorithms fail, and GAs should be applied to these problems.

Determining AoAs and allocation schemes to fit those AoAs has largely been a human-intensive product of wargaming. The MDMP relies on experienced soldiers to generate potential AoAs, and to consider during that process a wide variety of factors on the battlefield—from topography, weather, and time, to enemy capabilities, to the potential for surprise. Developing a robust way to automate and quicken this generation and pick out the few potential good AoAs from a sea of bad solutions reduces the effort expended by soldiers in the field. Because of the number of variables involved in choosing a good AoA, the method adopted must be

---

[5] The remainder of this report assumes a moderate background in basic GA techniques. For additional general information on military decisionmaking that also includes information on GAs, see Jaiswal (1997). For an introduction to GAs, along with the history and motivations behind their use, see Mitchell (1996).

good at finding those rare solutions among the many variables. GAs have been applied to such searching problems in the past and thus may provide attributes that facilitate such searching.

Perhaps the greatest advantage of GAs is that their design requires very few heuristics. Much like NNs and unlike the other algorithms previously mentioned, GAs discover the rules that create good solutions, and these rules are often ones that humans would rarely consider. The GAs' advantage over NNs is that their input and output design is highly configurable and more intuitive. NN input must be in a vector format, and certain input configurations may be better than others. NN outputs must be in a vector format and must be numbers between 0 and 1 or –1 and 1. Using heuristics, the designer must convert solutions and input data into a format that may not be either intuitive or optimal. GAs, on the other hand, merely need the input to be defined as the parameters of a fitness function whose output is a single number. The fitness function has an intuitive interpretation: It describes how "good" a solution is.

We have chosen to use a GA because it is fast, flexible, relatively intuitive and transparent, and lends itself to the discovery of a variety of options. A GA begins with a seed population of trial solutions and then evolves this population over several generations to find better and better solutions. The process is analogous to natural selection: Solutions are grouped by similarity ("niched"), combined to form new possibilities ("mated"), varied slightly to allow for incremental improvement ("mutated"), and finally evaluated against each other ("competed") to find the best of each generation to pass to the next. This process can be repeated a fixed number of times or until the solutions stop improving appreciably.

Model Overview

The purpose of this combat planning model is to develop options for Blue given both his mission and his intelligence about Red. As shown in Figure 1.2, this model finds several AoAs for Blue, allocates forces along these AoAs, and models the effects that Blue expects to encounter

Figure 1.2
Model Inputs and Outputs

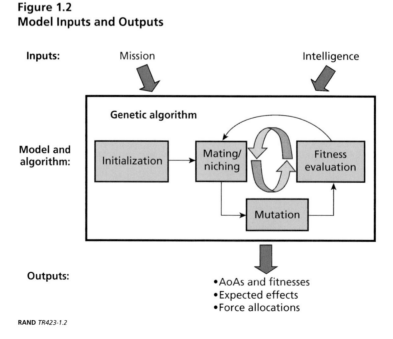

RAND *TR423-1.2*

along his routes given his intelligence about Red. Due to the large number of variables involved in the decisionmaking process, the "space" of Blue's possibilities is sufficiently vast that an exhaustive search within a reasonable time (preferably a few minutes) is currently impossible. Instead of employing brute-force computation, we need a means of exploring possible alternatives in a more intelligent and efficient manner; hence, we have chosen to use a GA.

As shown in Figure 1.3, our combat planning model is broken down into two phases, each of which uses a GA at its core. In the first phase, a GA finds a set of possible AoAs. In the second phase, a GA considers the allocation of Blue forces along these routes. We break the model into two phases, as opposed to tackling the entire problem in a single GA, to reduce computation time and hence improve the efficiency with which we find routes and force allocations. By breaking the problem into two halves and fixing the solution of the first half before solving the second, we have reduced computation time, accomplishing this through the elimination of part of our search space; we acknowledge that this comes at the expense of finding an optimal solution. However, we believe that this choice of algorithm is a reasonable compromise between computation time and optimality for two reasons: First, we were not guaranteed to find an optimal solution, given the nonconvex nature of the problem. Second, the process of first determining AoAs and then finding associated force allocations closely parallels the human decisionmaking process.

Although the model's outputs—a set of potential AoAs for Blue and an allocation of Blue forces to these AoAs—are concrete, the inputs—Blue's mission and Blue intelligence about Red—are largely conceptual. We use four key parameters to describe Blue intelligence about Red: intent, activity, capability, and location. Of these four, the first three are conceptual, and

Figure 1.3
The Model's Two Phases

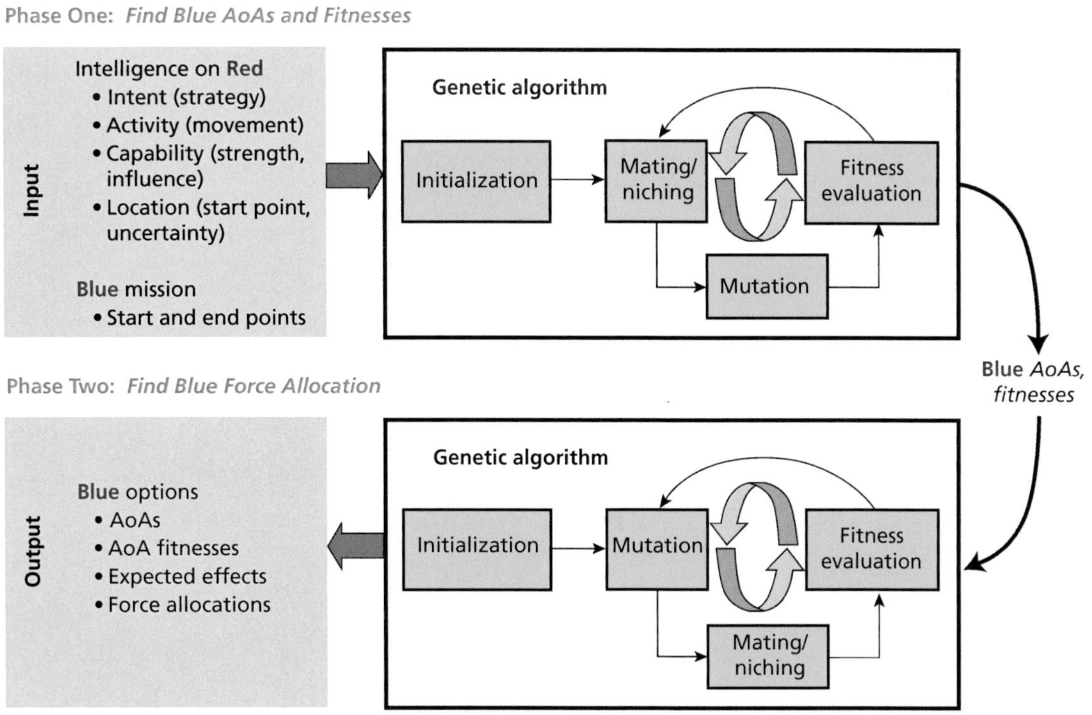

only the last is concretely definable. For the conceptual parameters, we describe a model that quantifies them and hence parameterizes the battlefield. Our methodology is mathematically complex, but this is appropriate given the complexity of a combat engagement. Still, under lying the mathematics, the model is at heart a combination of a few relatively simple ideas.

This report describes the quantification of the inputs, the process behind each of the model phases, and how the model illustrates the value of intelligence. In Chapter Two, we describe the modeling of enemy capabilities and the effects of enemy capabilities on friendly forces. In Chapters Three and Four we describe the model's first and second phases, i.e., the generation of AoAs and the allocation of forces to them. We introduce the terrain model in Chapter Five and explain how terrain affects both friendly and enemy movement. Chapter Six provides simulation results. In Chapter Seven we summarize our findings and present possible extensions.

CHAPTER TWO

Modeling Enemy Capability and Effects

This chapter describes a quantitative model of enemy capability and how this capability translates into an effect that Red can have on Blue. We visualize enemy capability with a topographical effect map. Figure 2.1 illustrates how a physical map that depicts the geographic location of various units can translate into an effect map that depicts the capability of those units at any point on the physical map.

In our model, the effect map illustrates the "effect" a Red unit can exert on a Blue unit at a given fixed position at a given point in time. In other words, a Blue unit views a Red unit as having an influence or "effect potential" on Blue's surroundings due to Red's various capabilities. The effect map is a generic picture of the capabilities of a Red unit that disregards the type of effect, which can be kinetic, nonkinetic, or both. We expect the Red unit's peak on the effect map to increase as that unit's capability increases.

Below, we introduce the concept of expected effect, which incorporates the uncertainty that exists in Blue's knowledge of Red's location. We then describe critical mathematical properties of the effect function, provide candidate functions, and present the effect function used in the model. Next, we derive an explicit expression for the expected effect using this effect function. To conclude, we relate the expected effect that Blue will likely encounter along his route to the fitness of the Blue route.

Figure 2.1
Enemy Capability and the Effect Map

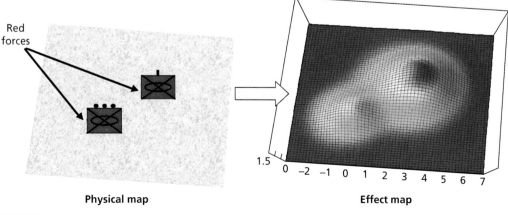

Expected Effect

The effect map construct, which describes Red's effect on Blue at a given fixed position and time, assumes that Blue knows Red's exact location. Of course, this is rarely the case. Instead, Blue typically possesses only *some* knowledge of Red's whereabouts. Using this knowledge, we describe an expected effect that incorporates the uncertainty in Blue's knowledge of Red's location (see Figure 2.2).

Figure 2.2
An Intuitive Understanding of Expected Effect

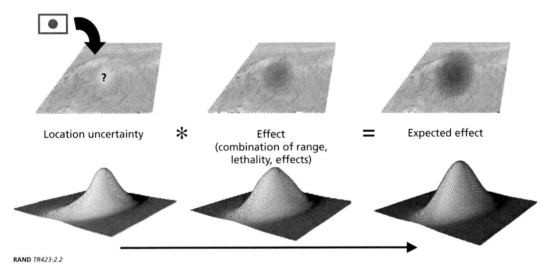

RAND *TR423-2.2*

The effect function, as we see below, depends on the distance between Red and Blue units. Because Blue is somewhat uncertain about Red's position, Blue cannot calculate the exact effect Red would have on it, but can only calculate an expected effect; this effect is averaged over the likelihood that Red is located in any given position.

We represent Blue's guess about Red's position by a probability distribution that describes the likelihood that Red is located at some distance from a best estimate. The location uncertainty and the effect function are convolved together to produce an *expected effect*. As seen in Figure 2.3, this convolution broadens and inflates Blue's estimation of Red's capabilities. This is due to uncertainty about Red's location.

Figure 2.3
Expected Effect Combines Location Uncertainty and Effect

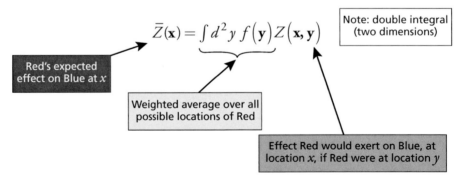

RAND *TR423-2.3*

The expected effect is simply the sum of effects that Red would exert on Blue at location **x**, at all possible locations of Red, **y,** weighted by the probability that Red is at **x**. In other words, it is the average effect. Although we assume that Blue has perfect knowledge of his own location, we have only a probabilistic description of where Blue thinks Red might be: the location uncertainty function $f(\mathbf{y})$. The double integral shown in Figure 2.3 is a mathematical representation of this averaging over the plane.

Here we assume that the area of engagement is small enough that the curvature of the Earth is unimportant. Although the horizon certainly affects Red's line-of-sight capabilities, this limitation in distance could be accounted for in the effect function. The incorporation of terrain in general is discussed in Chapter Five.

The Effect Function

A robust model should depend only on the general attributes of the effect function and not on its exact shape. Since the effect function is simply a mathematical approximation, sensitive dependence on its form would not be reasonable. This fact, however, gives us freedom to choose a mathematical function based on convenience. Our only requirement, then, is that the effect function satisfies a few basic properties.

First, the function should depend on the *distance* between Blue and Red units. It should be a symmetrical function with no preferred direction. In other words,

$$Z(x, y) = Z(\| x - y \|) = Z(r),$$

where x and y are vectors that describe Blue and Red locations, respectively, and r is a scalar that describes the distance between them. The incorporation of terrain effects, described in a later chapter, does not alter this property.

Second, the effect function should be *additive*, meaning that the effect of two Red units is the sum of the individual effects of each, and the effect is cumulative over time. This property permits the straightforward calculation of massing effects. Additivity is not a trivial property. Indeed, military units arguably experience increasingly larger benefits as they mass together. We start with an additive function not because we chose to ignore these synergistic effects, but because we show how to incorporate them later.

Finally, a minimum number of *parameters* should shape the function, and these parameters should be easily related to the underlying capabilities of the Red unit. In the cases described in this report, capabilities are parameterized by the unit's "strength" and "range," both of which can be deduced from quantitative values calculated from weapons effectiveness scores or similar measurements of military strength.

Mathematically speaking, the Red effect function should

- Be nonnegative and finite valued everywhere, and have a finite integral over all space. Specifically,

$$0 \le Z(r) \le Z_0, \int Z(r) d^2 r = a$$

should hold true for some positive finite values Z_0 and a.
- Be smooth, continuous, and differentiable. This property proves important when we describe how the effect function is used to quantify route fitness.
- Decrease monotonically as the distance r between the Blue and Red units increases, attaining its maximum value at the origin ($r = 0$). Specifically,

$$\frac{\partial Z(r)}{\partial r} < 0, r > 0$$

should hold true.
- Have an inflection point, or knee in the curve, at some finite distance r_0. This inflection point is required to fix and control the effective range of the Red force. Specifically,

$$\left. \frac{\partial^2 Z(r)}{\partial r^2} \right|_{r=r_0>0} = 0$$

should hold true.

Many functions meet these criteria, but some of the simplest do not. We evaluated several candidate effect functions (see Table 2.1).

Power Law. The first type of function we considered is a power law. After all, almost every force in the known universe falls off as a power law (often an inverse square law). However, this function does not satisfy our three criteria. Not only is the function ill behaved at its center,

Table 2.1
Candidate Effect Functions

Candidate	Example	Evaluation Criteria	Comment
Power law	$Z(r) = \dfrac{1}{r^\alpha}$	Finite integral: Fails (infinite valued at origin) Decreasing: Satisfies Knee in curve: Fails (has no inflection point)	Fails too many criteria
Hyperbolic function	$Z(r) = 1 - \tanh^\alpha(r)$	Finite integral: Satisfies (for $\alpha \ge 1$) Decreasing: Satisfies (for $\alpha \ge 2$) Knee in curve: Satisfies (for $\alpha \ge 1$)	Difficult to work with
Gaussian	$Z(r) = \dfrac{1}{2\pi\lambda^2} e^{-(r^2/2\lambda^2)}$	Finite integral: Satisfies Decreasing: Satisfies Knee in curve: Satisfies	Well understood and simple

a point we cannot ignore, but it does not have a knee in the curve. Either it falls off too fast, giving the Red unit almost no range to speak of, or it falls off too slowly, giving Red nearly infinite range.

Hyperbolic Function. Trigonometric or circular functions (e.g., sine, cosine) decrease favorably; however, they oscillate and are therefore clearly unsuitable. The nonoscillatory version of the circular function is the hyperbolic function (e.g., the hyperbolic tangent), which does, in fact, possess all the desired properties. Unfortunately, this function is not easy to work with.

Gaussian Function. For the effect function, we have chosen to use the Gaussian function, also know as "the bell curve" or "normal distribution." It is well understood, it fits all three criteria, and it is mathematically easy to work with. It also happens to be the same shape as the location uncertainty function, which is traditionally a normal distribution.

Explicit Expression for Expected Effect

Having now selected a specific effect function, we can compute the expected effect. Here, the power of the Gaussian choice is clear: We can perform the integral over all space (i.e., we can conduct the averaging) explicitly and arrive at a single function that combines both effect and uncertainty. Not surprisingly, this new function is also Gaussian.

The location uncertainty is parameterized by the usual standard deviation (σ). It describes an error circle that surrounds the expected location of the Red unit. Although we could have chosen an error ellipse, breaking symmetry would have added a great deal of complexity for little reward. The uncertainty in Red's position is represented as a two-dimensional normal distribution:

$$f(y) = \frac{1}{2\pi\sigma^2} e^{-\frac{|y-\bar{y}|^2}{2\sigma^2}} .$$

The standard deviation of the effect function, λ, does not represent an error (as it does in the location uncertainty function), but rather a range. It represents the knee in the curve. The only remaining parameter is the height of the Gaussian function. In this case, we use the ratio of the strengths of the Red and Blue forces (R/B). The effect function, $Z(\boldsymbol{x},\boldsymbol{y})$, is represented by a two-dimensional Gaussian function in distance between Red and Blue, in which $\mathbf{x}$ and $\mathbf{y}$ are position vectors:

$$Z(x,y) = \left(\frac{R}{B}\right) \frac{1}{2\pi\lambda^2} e^{-\frac{|x-y|^2}{2\lambda^2}} .$$

R/B is *not* the numerical ratio of the number of forces, but rather a ratio of their strengths designed to measure capabilities. The ratio provides an estimate of the relative strength of Red and Blue forces, should they come into direct and immediate engagement. These two parameters can be estimated from knowledge of the range and lethality of the weaponry of each unit.

Various scales have been developed to incorporate capabilities into quantitative combat simulations and analyses as well as to support disarmament and strategic level force-balancing policies. Well-known systems of scoring—such as the weapon-effectiveness index scores used in combat simulations (Murray, 2002), UK variants such as Balance Analysis & Modeling System scores, and variants of these methods (for example, see Allen [1992])—should be evaluated and adopted in accordance with the effects that are most important to the decisionmaker.

The convolution of the effect and uncertainty in location, known as the expected effect, is again a two-dimensional Gaussian function,

$$\bar{Z}(x) = \left(\frac{R}{B}\right) \frac{1}{2\pi(\sigma^2 + \lambda^2)} e^{-\frac{|x - \bar{y}|^2}{2(\sigma^2 + \lambda^2)}}$$

whose standard deviation is

$$\sqrt{\sigma^2 + \lambda^2}.$$

Relating Expected Effect to Route Fitness

So far, everything we have discussed concerning effect and uncertainty in location has related to a single snapshot in time. Before we proceed to the GA to discover Blue routes, we need to understand how to find the expected effect along a route. First, note that any route Blue takes will cut a path through the expected effect map as shown on the left side of Figure 2.4. We refer to the cross section that results as the "expected effect profile."

In Figure 2.4, we assume that Red is stationary, but in reality, of course, Red will move. For a moving Red, we need a four-dimensional or animated plot to show the changing effect map over time. The effect profile, however, is still a two-dimensional plot of expected effect

Figure 2.4
Expected Effect Map and Profile

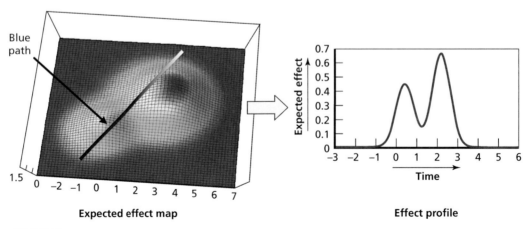

Expected effect map

Effect profile

over time that will resemble the chart on the right side of Figure 2.4. Indeed, Blue will have to consider many possible movements of Red, including an adaptive Red that reacts to Blue's behavior. We discuss such behavior later when we describe the GA itself.

We use measures of the effect profile to quantify the attributes of a Blue route. The selection of such measures to define the suitability or fitness of a Blue route is a separate matter, and there are many such measures, as we discuss below. We chose not to look for an absolute number that measures the fitness of a route across scenarios, but rather for a means of comparing two routes for the same scenario. As such, comparisons of specific fitnesses across scenarios are not possible with our model, but this utility is not a necessary characteristic of a tool designed to facilitate decisionmaking within a given scenario.

Summary

This chapter describes a quantitative model of Red capability and how this capability translates into a potential effect on Blue. We first defined a topographical effect map that is based on a physical map of Red units. The effect map, a generic picture of the kinetic and nonkinetic capabilities of a Red unit, describes Red's effect on Blue at a given fixed position at a given point in time. Since Blue is unlikely to ever know Red's exact location, we convolved this effect map with a location uncertainty function to obtain the expected effect (which is represented by another topographical map).

Next, we described desired mathematical criteria for the effect function and summarized our evaluation of three candidate functions. We explained our decision to model the effect function as a Gaussian function, which we also use to model the location uncertainty function. From the effect and uncertainty functions, we derived an explicit expression for expected effect. This new function, the convolution of effect with location uncertainty, is the convolution of two Gaussians; unsurprisingly, it is yet another Gaussian function.

We have discussed Red capability and location, but have not yet addressed intent and activity. Those parameters are described later, since they are scenario specific. We are now ready to discuss the GA, which will discover Blue routes on a battlefield of moving enemies.

CHAPTER THREE

Generating Blue AoAs: The Phase One Genetic Algorithm

The first phase of our model uses a GA to evolve a population of potential Blue routes. As outlined in Figure 1.3, the GA begins with an initialization phase and then iterates through the processes of niching, mating, mutating, and evaluating fitnesses. We discuss each phase in turn.

Initialization

During the initialization phase, we generate a seed population of potential routes. These routes are generated randomly as follows: Each route begins as a straight line between the start and destination points and is then decorated with a random number of "waypoints" whose total number is drawn from an exponential distribution. Waypoints are not positioned on the straight-line path, but rather are placed somewhat off route to produce a more diverse set of paths. To incorporate a waypoint, one of the route segments is split into two smaller segments, bending outward or inward as necessary to include the waypoint. To insert a waypoint, we randomly choose a location along one of the route segments, and then add a new point perpendicular to this segment at some distance in either direction. This distance is drawn from a normal distribution whose mean is zero and whose standard deviation equals half of the original segment length.

Mating and Niching

Once the initial population of routes is established, we pair off the routes and mate them to produce two new "children" routes. Every route is used only once as a parent and mates with only one other route. We also apply "niching" rules to limit the possible pairings. (These mating and niching procedures are discussed in detail later on in this report.) As in asexual reproduction, each generation's unmated routes produce one child, an exact clone, except when mutations occur.

The mutation step is important because it helps the GA discover new and interesting routes. We implement mutation by allowing routes to add new waypoints or remove old ones. After the mating procedure has produced children, the children are mutated with some probability.

After mating and mutation, we have many families of four (two parents and two children), plus a few "single-parent families" that consist of an unmated route and its mutated child. The children and parents then compete against each other within each family, and only the two fittest routes (or single route, in the case of the single-parent family) survive and are passed on to the next generation. By forcing such competition, we ensure that the minimum (and maximum and average) fitness of all routes in each generation monotonically increases with each new generation.

Next, we explain why the niching process is an important component of the GA. Then, we describe the details of the niching algorithm used in our model. We then discuss how routes are mated and mutated. Next, we address the Red behavior model, which describes how Blue anticipates Red movement. We conclude with a description of route fitness, which ultimately determines survival in the GA.

The Need for Niching

The purpose of niching is to maintain diversity among a population of potential solutions. Since the purpose of the first phase of this model is to find a range of AoAs for Blue, and not merely a single best path, we need to explore the space of potential Blue routes and allow routes to settle into local optima. To find these local optima, known as "niches," with a GA, you must mate like with like. This type of mating effectively breeds for unusual traits. Without niching, it is primarily the dominant traits that are passed on, and in time, the population is likely to approach homogeneity (Figure 3.1).

Just as the mating and mutation steps in the GA mimic a biological process, so does the niching procedure. In nature, speciation is a niching process that allows different types of organisms to develop in a common environment. Different species survive by exploiting their environment in unique ways. Niching aids in the discovery of more than one solution to a problem because it prevents dissimilar organisms from mating. Once two organisms are too different to mate, it is unlikely that their offspring will be compatible. Thus, species are born. In nature, members of different species rarely mate, and hybrids, when they do occur, are usually infertile (e.g., mules).

One of the nice properties of GAs is the transparent manner in which niching can be accomplished. Because GAs use mating to combine potential solutions to create new ones, niching is required to keep apart those "genomes" that solve the problem in radically different ways, ways that when combined, potentially yield no solution at all. Due to the niching algorithm, the GA presents us at completion with several possible solutions, one for each niche.

Figure 3.1
Over Time, Mating Without Niching Can Result in Homogeneous Paths

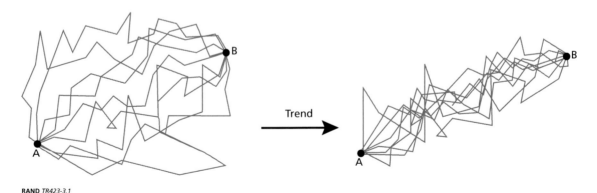

These solutions may not be equally fit, but may be quite different, and each may have appealing qualities. Furthermore, when GAs attempt to solve hard problems with large state spaces, niching can be crucial to the discovery of any satisfactory solution.

Consider the following example: An infantry unit is searching for a path around a mountain range. There are two paths the unit might take, one on each side of the mountain. In other words, there are two possible solutions to the problem. Without niching, a GA might attempt to "mate" these two solutions and eventually obtain an "averaged" path that goes directly over the mountain peak—an unacceptable alternative that may not preserve either of the two parent paths. However, niching prevents these parent paths, which are viable solutions, from becoming mates; thus, their niches are preserved.

The Niching Algorithm

Nature has its own means of determining when organisms are compatible. We must use a metric to tell our algorithm when two species are in two separate niches and hence not allowed to mate. There is no simple way to compare two paths and assign a number to the "difference" between them. Our method is a compromise between optimality and runtime. It extracts a small set of parameters from each path and compares this set only, rather than comparing entire paths. Although this method discards some information, it speeds up the GA significantly. We also search for potential mates by random draw rather than exhaustive search. In other words, we look for a mate randomly and, if we find one within a set number of trials, we proceed. Otherwise, we declare the path to be an unmated bachelor and move on. The persistence with which we search for a mate can greatly affect the runtime.

One important attribute of our niching algorithm is that it is scenario independent. We can select any two paths, compare them, and determine whether they belong to the same niche or not based on a criterion that is internal to the model and independent of the scenario.

For the purpose of niching, we characterize routes by their centroids, or average positions, as shown in Figure 3.2. To determine if two routes are suitable mates, we first measure the "distance" between them by measuring the distance between their centroids; then, we determine if this distance lies within a threshold limit. We must be careful that the niche threshold we choose does not make the algorithm scenario dependent. Hence, we cannot simply pick a fixed threshold, since the size of niches cannot be known ahead of time. Instead, we select a threshold based on the nature of the population itself. If the population is widely spread, we

**Figure 3.2
Paths Are Parameterized by Centroid**

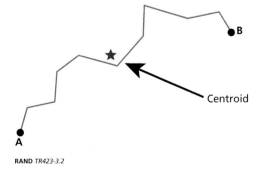

look for larger niches by being less restrictive in our mate selection. However, if the population is narrowly distributed, we adjust our niche threshold accordingly, becoming more restrictive.

To determine how "spread out" a population is, we look at the standard statistics: the mean, variance, and covariance of the set of centroids. We use these statistics to define a covariance ellipse, allowing paths to mate only if their centroids fall within this ellipse. Specifically, given the covariance matrix **S** of the set of centroids, paths with centroids **u** and **v** are suitable mates if

$$\sqrt{(u - v) \cdot S^{-1} \cdot (u - v)^T} \leq 1.$$

The centroid of every path in the population is located and extracted, as shown in Figure 3.3. From that set, the mean, variance, and covariance are calculated. The distance between the centroids (normalized by the standard deviations in each direction) is used as a measure of the dissimilarity of the paths. Only paths whose centroids are suitably close (i.e., are within the covariance ellipse shown in Figure 3.4) are mated. This method is further explained in our description of the model's second phase.

Figure 3.3
Centroids Are Extracted for Every Path in the Population

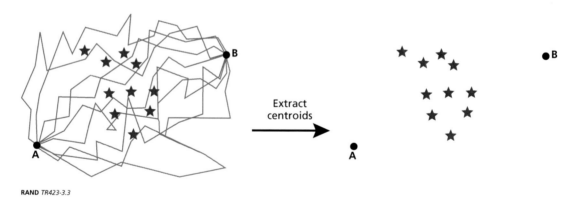

Extract
centroids

Figure 3.4
Paths Whose Centroids Lie Within the Covariance Ellipse Are Potential Mates

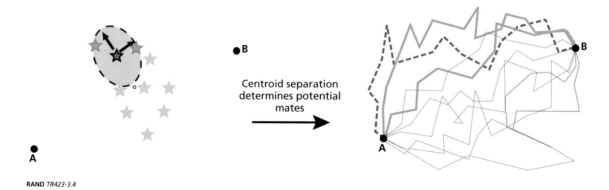

Centroid separation
determines potential
mates

The Mating Procedure

The first part of the evolution process is mating. We use a standard crossover procedure to generate two children, as shown in Figure 3.5. First, a crossover point is randomly selected from the intermediate waypoints of the parent path that possesses the fewest waypoints. For example, if Parent 1 has five waypoints and Parent 2 has seven, then the crossover point is randomly selected from waypoints two, three, and four. The crossover point cannot be either of the endpoints, namely waypoints one and five. If waypoint three is selected as the crossover point, the two children paths are constructed as follows: One child path consists of the route segments that join the starting point (in our case it is the same for all paths) to the crossover point on Parent 1, namely the segments joining waypoints one to three, followed by the route segments joining waypoint four on Parent 2 to the endpoint. The second child is simply the complement: It consists of the route segments that join the starting point to waypoint three on Parent 2, followed by the segments joining waypoint four to the endpoint on Parent 1.

Since the point at which the crossover occurs is selected randomly, it may be closer to one of the endpoints than the other. Because we impose no requirements on the proximity of crossover points (i.e., waypoint three in parent paths 1 and 2 in the example above), child paths can make a sudden jump at the point of crossover. Although this procedure can create jagged paths, such paths are likely to be considerably less competitive or fit than their parents.

All four routes compete against each other, and only the two fittest are passed to the next generation. Before that competition occurs, however, each child route mutates slightly with some probability.

Figure 3.5
Route Mating Procedure

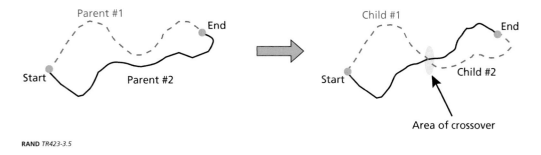

Mutation

As illustrated in Figure 3.6, we mutate routes in one of two possible ways: by adding a new waypoint or by subtracting an old one. We can choose either of these two methods with some probability, or choose not to mutate at all. We bias the simulation at least two to one in favor of adding additional points to make the routes more complex. While it often helps to add waypoints more than to subtract them to search the space, this addition of points also makes the route more circuitous. Hence, we also allow the option of subtracting a waypoint. If we did not do this, it would be difficult for a route to "straighten out" under the pressure of constant mutations.

Figure 3.6
Route Mutation Procedure

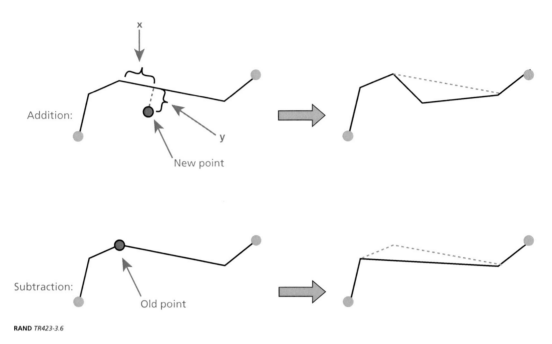

RAND *TR423-3.6*

In the waypoint addition mode, each segment is weighted by its length, and the segment to be mutated is selected randomly. Hence, longer segments are more likely to be broken. The distance along the segment, x, is chosen randomly from a uniform distribution. The perpendicular displacement from the chosen segment, y, is taken from a normal distribution whose mean is zero and whose standard deviation is half the original segment length. In subtraction mode the segment lengths are not weighted, but the point to be removed is still chosen randomly. Endpoints cannot be removed.

After the routes have paired, mated, and produced offspring, and those offspring have been mutated, routes are then compared for fitness. We use the expected effect profile of the Blue route as the basis for this calculation, as discussed earlier. First, however, we must address Blue's perception of Red and Red's resulting motion model.

Red Behavior Model

Since the calculation of the effect profile of a Blue route necessarily assumes knowledge or inference of the average or expected position of Red, we must describe our model of how Blue anticipates Red movement. We cannot, of course, simply use the ground truth of Red's position, for that would give Blue too much information. The model of Red movement, or more accurately, Blue's perception of Red movement, plays a crucial role in the development of Blue routes. This model also enables us to explicitly incorporate Blue's perceptions of Red's strategic intent, and Red's intelligence and adaptability.

Red Intent

To determine Red behavior, we first need to quantify what Blue knows of Red's intent and what "intent" means in our model. We allow Red to take one of several possible strategies, and we assign a fixed likelihood to each, representing Blue's belief that Red will take that strategy. The possible strategies are

1. **Hold ground.** Red stays where he is, regardless of what Blue does.
2. **Move to location.** All Red units head directly to a specific location, without regard to (or knowledge of) Blue's mission.
3. **Individual intercept.** Each Red unit attempts to intercept Blue's route as quickly as possible, without coordinating with other Red units.
4. **Coordinated intercept (or mass effects).** All Red units attempt to intercept Blue's route at the same point. This strategy requires coordination among the Red units. We assume that each Red unit has perfect knowledge of all other Red unit locations and hence can coordinate his efforts, but that Red may have imperfect knowledge about Blue's location at any given point in time.
5. **Attempt to escape.** All Red units flee from Blue as quickly as possible.

Since the focus of this model is to understand planning, route selection, and allocation, Blue calculates the effect profiles assuming that Red experiences no loss of will or attrition. The incorporation of expected loss from attrition or loss of will can occur in the evaluation of the fitness of the effect profile. Also note that although these potential Red strategies are not absolute and orthogonal—indeed, Red might be expected to assume varying degrees of each type of strategy for individual units—these five potential Red strategies are assumed to be *representative* of the landscape of possible Red strategies. In fact, some of the strategies, for instance mass effects, are worst-case scenarios because they assume that Red's self-knowledge is very good. These demanding scenarios are appropriate for the purposes of planning, since Blue should plan against worst-case scenarios for each strategy rather than against less stressful but potentially more-plausible scenarios.

Although Blue is aware of the five possible strategies that Red may adopt, Blue does not have complete knowledge of what Red will do. We parameterize Blue's knowledge of Red's intent by assigning a set of normalized probability weights, w, to each strategy. These weights indicate Blue's belief that Red will follow a particular strategy. Thus,

$$w = \left(w_1, w_2, w_3, \ldots w_n \right).$$

For all i,

$$w_i \geq 0$$

and

$$\sum_{k=1}^{n} w_k = 1.$$

For example, the kth strategy has an assigned probability of w_k, which indicates that Blue believes Red will follow strategy k with probability w_k.

We do not update Blue's knowledge of Red intent as the scenario progresses. The model is intended to be a planning tool and, as such, Blue must plan based on his best available knowledge of Red intent.[1] Thus, while planning, Blue uses only the information at hand and does not model the possibility that learning more of Red's intent in the future will allow him to improve his own choices.

Although a set of n weights is a good description of the knowledge of Red intent, we prefer a single statistic that measures this knowledge. The variance of this set of weights is such a statistic, and it is a fairly intuitive measure of Blue's knowledge of Red intent. If all weights are equal, Blue knows nothing, and the variance is correspondingly 0. On the other hand, if one weight is 1 and the rest are 0, then Blue is certain of Red's intentions, and the variance is in fact maximized.

A little exploration shows how to normalize this into a 0 to 1 measure.[2] In our method, we use the statistic W, where

$$W \equiv \frac{n^2}{n-1} \times \sigma_w^{\ 2}$$

and $\sigma_w^{\ 2}$ is the variance of w. In the limit of a large number of strategies, if the range of possible Red strategies is effectively narrowed down to 1, then W is 1; if it is narrowed down to just two choices (equally weighted), then W is one-half; if it is narrowed down three choices, W is one-third, etc. We use the W statistic as a measure of Blue's knowledge of Red intent.

Red Intelligence and Adaptability

With a handle on Blue's knowledge of Red intent, we can now model (Blue's perception of) how Red will react to Blue's movement. For each potential Red strategy, we model Blue's perception of Red movement. For the most basic strategies, the first (hold ground) and the second (move to location), Red is assumed to be nonreactive. In other words, Red's path is preplanned, and hence calculating the effect profile on any given Blue route is straightforward.

For the remaining strategies, we assume that Red will react to what he knows about Blue, and to what he believes he can predict about Blue's future behavior. The path Red takes will depend on Red's intelligence and adaptability.

To determine Red behavior, we slowly reveal Blue's path to Red, allowing Red to adjust his path accordingly. We measure intelligence and adaptability in terms of time steps of knowledge, which are the total number of time steps of Blue's path that have been revealed to Red at any given time.

There are three key parameters in the Red intelligence model: $\boldsymbol{L_0}$, ΔU, and D. At the start of the model (t = 0), we give Red a head start of $\boldsymbol{L_0}$ time steps of Blue movement. This represents

---

[1] In a real situation, if a commander learns additional information that might help in planning, the model can be run with an updated set of inputs. However, the model will still be a snapshot based on best available information.

[2] The desire for a 0 to 1 measure for Blue's understanding of Red intent is driven by previous RAND research (Pernin et al. [2007]) that explored metrics for incorporating quantitative fusion algorithms into constructive simulations. In that research, location, activity, capability, intent, and other knowledge types are incorporated as probability distributions that relate qualitative measures to quantitative metrics for manipulation. Obtaining the data to support the method described above would require such a scheme.

Red's initial intelligence about Blue's activity. L_0 may also be negative, indicating that Red does not realize Blue has begun to move until $|L_0|$ time steps have already passed.

As time progresses, we allow Red to receive periodic updates. The update interval, ΔU, represents the number of time steps between intelligence updates, thus setting the size of the decision loop. This parameter reflects Red's adaptability.

Parameter D is the rate at which Red's intelligence is gaining on Blue or falling behind. The size of the update matters less than how that size compares to the update rate. If D > 1, Red has good intelligence and will eventually identify Blue's full route before Blue reaches his destination. If D = 1, Red's intelligence is good enough; Red will be able to keep up with Blue's movement sufficiently to maintain his intelligence advantage (or disadvantage), staying that many steps ahead (or behind). If D < 1, Red's intelligence is poor; Blue will reach his destination well before Red realizes where Blue is going. If Red is falling behind, Red will try to project Blue's movement into the future until the time of the next update (t + ΔU).

Determining Route Fitness

As previously mentioned, we use the expected effect profile of each route as the basis for assessing route fitness. Now that we have a model of Red behavior, we can determine an anticipated Red route; from this route and the corresponding Blue route, we can determine the expected effect profile for Blue.

The selection of a metric for the expected effect profile is a matter of choice. Note that Blue's objective should be reflected in the calculation of route fitness. Since we are considering a case in which Blue aims to minimize his exposure to Red, we chose a metric that minimizes some measure of the expected effect. Specifically, we use the sum of the effect over all time, or the "total effect," as the single key parameter for each route, as shown in Figure 3.7.

Figure 3.7
Example Fitness Metric—Calculation of the Area Under the Curve

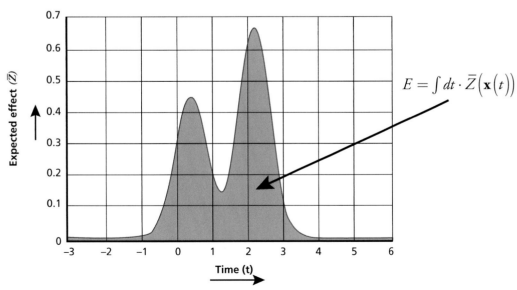

$$E = \int dt \cdot \overline{Z}\left(\mathbf{x}(t)\right)$$

We could, however, consider the maximum effect attained as the key parameter,[3] or attempt to count the number of peaks above a threshold. Each choice leads to considerably different results for AoA selection. The choice of evaluation criterion should reflect the interests of the commander or subject matter expert running the model.

The next step is to relate the total effect to route fitness. We desire a fitness function that varies inversely with total effect and that runs between 0 and 1 for easy normalization (i.e., it has no negative values and it never runs away to infinity). We also desire an additional tuning parameter that will allow us to adjust Blue's willingness to encounter Red's effect. We have chosen for our model of route fitness, f, a decaying exponential function whose exponent equals the total effect scaled by $(1/\varepsilon)$, as shown in Figure 3.8. Thus,

$$f \equiv e^{-E/\varepsilon}.$$

Figure 3.8
The Fitness Function

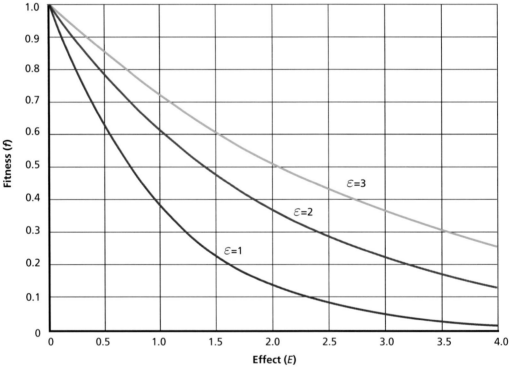

RAND *TR423-3.8*

---

[3]   In the case of strict equivalent force size comparisons, one might consider using the 3-to-1 rule, applying a large penalty to those paths that violate it. In these cases, however, the value that comes from the calculation of the entire profile may be lost. See Davis (1995) for a more complete discussion.

The parameter ε represents Blue's willingness to engage, or in other words, Blue's willingness to expose himself to Red. As ε is decreased, the fitness function becomes more sensitive to changes in the total effect, which translates into greater penalties for Blue (i.e., lower route fitness) due to an unfavorable force ratio or to prolonged proximity to Red or both.

To incorporate Blue's knowledge of Red intent, we need to compute the basic fitness of the route, f, under the assumption that Red could follow each potential strategy. We then compute the weighted sum of these fitnesses, each weighted by Blue's belief that Red will follow that strategy. The result is the strategic fitness, $\bar{f}$, expressed as

$$\bar{f} \equiv \sum_{k} w_{k} f_{k},$$

where each of the basic fitnesses (f_{k}) is computed separately, assuming that Red follows the corresponding kth strategy.

Summary

The first phase of the GA discovers Blue routes by initializing a population of those AoAs and then iterating through the processes of niching, mating, mutation, and fitness evaluation. The Blue route is defined by a list of waypoints that join the start and end points of the route. To initialize the first population, we begin with straight-line paths between the start and end points and then decorate these paths with randomly assigned waypoints whose total number is drawn from an exponential distribution. To determine suitable mates, we use a niching algorithm that (1) parameterizes each route by its centroid and then (2) determines the Mahalanobis distance between pairs of centroids to assess whether two routes are similar enough to mate. When two parent routes are found, they are mated by a crossover procedure, and their resulting children are mutated. We next determine the fitnesses of the resulting child routes to assess which will survive into the next generation. Child routes must compete with their parents, and only the top two fittest routes in each four-member family of AoAs survives.

To determine the fitness of a route, we first consider how Blue anticipates Red's movement; this movement is described by a Red behavior model. We begin by parameterizing Blue's knowledge of Red's intent by the probability that Red will take each of five distinct strategies. Then, for each of those strategies that depends on Blue's route, we define a Red route based on a model of Blue's perception of Red's intelligence and adaptability. Given the route that Blue expects Red to take, we can determine an expected effect profile of Blue's route through Red's topographical expected effect map. We use the total effect, or area under the expected effect profile, as the key parameter for determining route fitness. Route fitness is then defined by a decaying exponential function with an exponent equal to the total effect, scaled by $(1/\varepsilon)$. We then compute this fitness for each of the possible Red strategies, weight each by the probability that Red takes the corresponding strategy (this step represents Red's intent), and then sum the result. This result is the strategic fitness. Chapter Four describes the second phase of the GA, which generates Blue force allocations.

Generating Blue Allocations: The Phase Two Genetic Algorithm

The second phase of the model is designed to be similar to the first. Again, a GA forms the core of the phase, which consists of an iterative process of niching, mating, mutating, and competing among possible solutions. In the second phase, this process produces *Blue allocations*. A Blue allocation is a complete assignment of all Blue forces to the AoAs discovered in Phase One. In a Blue allocation, several or no Blue units can be assigned to a particular AoA. The iterative process used to evolve populations of Blue allocations is quite similar to that described in Phase One, which evolved populations of Blue AoAs. In Phase Two, every Blue allocation is mated exactly once, each mating pair produces exactly two children, and only the two fittest allocations from each "family" survive into the next generation.

Although the iterative process used in Phase Two closely resembles the one used in Phase One, there are some significant differences in the associated algorithm. Recall that in Phase One, when discovering Blue AoAs, we considered the case of a single Blue unit facing all of the Red forces. This is a worst-case scenario in which all of the Red forces were assigned to respond to a single Blue unit in accordance with one of the five strategies. Typically, Red responded by engaging Blue, but Red can also flee; in some cases, Red pursued his own plans without regard to Blue, resulting in no response at all. To determine the fitness of a Blue allocation, we can no longer assume that Blue faces the combined attention of all the Red forces. Instead, we must now describe how Red chooses to array his forces; we must quantify Red's *activity*.

To define Red's activity, we ask the following question: If Red observes Blue units pursuing a particular AoA, which Red units, if any, will respond (by engaging, fleeing, ignoring, or holding ground)? Here we simplify the situation by assuming that one Red unit cannot respond to advances along two AoAs simultaneously; instead, we presume that the Red unit will concentrate his efforts on one AoA only. We also assume that Red units will be assigned to deal specifically with advances along AoAs rather than be arrayed against specific Blue units. In other words, we assume that Red uses a "zone" defense rather than a unit-on-unit defense. However, we do not mean to imply that Red is acting defensively.

We then define Blue's knowledge of the activity of each individual Red unit as Blue's level of knowledge about with which AoA (which "zone") each Red unit will concern itself. This process is described in greater detail below. At the moment, we simply wish to observe that just as we speak of Blue allocations to the set of AoAs, so we can also speak of *Red allocations* to the same set.

Because Blue's knowledge of Red activity is unlikely to be perfect—in other words, Blue does not know with certainty how Red will allocate his forces—Blue must consider a *field* of possible Red allocations. To compute the fitness of a Blue allocation, we evaluate this allocation against a field of plausible allocations of Red forces, which are, as noted above, assigned to oppose or attack along the various AoAs. In the case of perfect knowledge of Red activity, there is by definition only one possible Red allocation, and hence our "field" reduces to a field of one.

Next we describe the niching algorithm used to find viable mates. We then discuss how Blue allocations are mated and mutated. Next, we describe how a field of Red allocations is defined based on Blue's knowledge of Red's activity. After explaining how to calculate the fitness of a Blue allocation, we relate this fitness back to the parameters that describe Blue's mission and intelligence.

The Niching Algorithm

Recall that during Phase One of our model, we applied niching to the centroids of candidate AoAs. In Phase Two, we apply niching to candidate force allocations, where each force allocation is represented by an n-dimensional vector (if there are n AoAs) with components that correspond to the total strength of units assigned to each AoA. Figure 4.1 graphically depicts eight distinct allocations to three AoAs. Hence, each allocation is a point in 3-dimensional space.

For example, assume there are three units of equal strength and an allocation of [1, 0, 2]. This allocation vector signifies that we have assigned one unit to AoA 1, none to AoA 2, and two to AoA 3. (If the units are of unequal strength, weight them accordingly.)

Given this representation of an allocation, we can use a niching algorithm similar to the one that described in Phase One. Recall that in Phase One, the points existed in two dimensions and represented the centers of mass of the candidate routes. In Phase Two, we have points in n dimensions that represent exactly a Blue allocation (as opposed to some measure of it).

To determine if two allocations are "close enough" to be mated (and hence belong to the same niche), we look at the distance between these points weighted by a measure of the variation of all points in the population. As in Phase One, we recognize that the population of

Figure 4.1
Allocating Forces Across Three AoAs

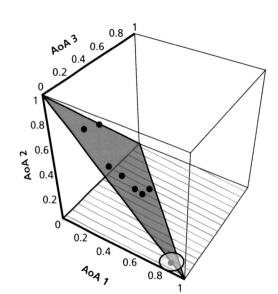

points may be more spread out in one direction than another. Hence, to find a distance metric that takes this variation into account, it is necessary to weigh the distance by the population variances in each direction. In the two-dimensional space considered in Phase One, we only needed to consider the covariance ellipse to find such a metric. In Phase Two, we are interested in higher dimensions, since the number of AoAs determines the dimension of the space. The generalization of this metric for higher dimensions is the well-known Mahalanobis distance, d, which weights the standard Euclidean distance by the inverse of the covariance matrix, **S**. For two points, **u** and **v**, in n-dimensional space, the Mahalanobis distance between them is given by

$$d\left(\mathbf{u},\mathbf{v}\right) \equiv \sqrt{(\mathbf{u}-\mathbf{v}) \cdot \mathbf{S}^{-1} \cdot (\mathbf{u}-\mathbf{v})^{\mathrm{T}}}\,.$$

All points equidistant from a common point using this metric lie on an ellipse. Hence, in two dimensions, the level curves of this metric are concentric ellipses.

Here we note an important implementation detail of the niching algorithm. Observe that the sum of all allocated forces, which is the sum of components in our force allocation vector, is always constant. Hence, all points in the "allocation space" lie in a plane, a subspace of dimension $n - 1$. Because of this degeneracy, the $n \times n$ covariance matrix will be singular (noninvertible).

Fortunately, we can eliminate the extra dimension manually if we rotate our coordinates into this subspace and then consider only the $(n - 1) \times (n - 1)$ covariance matrix of the (new) coordinates within the subspace. For two AoAs (two dimensions), this is the equivalent of "rotating" our coordinates to consider the sum and difference of the force strengths assigned to the two AoAs. Since the sum of all force strengths remains fixed, the (signed) difference between those assigned to the first and second AoAs is sufficient to completely specify the allocated strengths.

Hence, due to the singularity of the covariance matrix, we must rotate our coordinates before we compute the Mahalanobis distance. After the rotation, the resulting point in the "allocation space" will have n − 1 independent coordinates. We use these coordinates to build the covariance matrix required to evaluate Mahalanobis distances.

The Mating and Mutation Procedures

To mate and mutate strategies, we need a mathematical description of a Blue allocation that facilitates these processes. We represent a Blue allocation as an ordered list that contains precisely one entry for each individual Blue unit (as opposed to one entry for each AoA that was used in the niching algorithm). The entry for each unit shows which AoA that unit will pursue. For example, an allocation of [1, 3, 3] means the first unit takes AoA 1, while units two and three take AoA 3. In this example, no unit takes AoA 2.

To mate two force allocations, we apply the following rules:

1. If a given unit is assigned to parents with two different AoAs, it will be randomly assigned to one of those AoAs in one offspring and the other AoA in the other offspring.
2. If a given unit is assigned to parents with same AoA, it will be assigned to that same AoA in both offspring.

To decide which child draws from which parent, we flip a coin once for each unit in the allocation. For example, consider the case where Parent 1 is the Blue allocation [1, 2, 3] and Parent 2 is the Blue allocation [2, 1, 3]. Since both parents have placed the third unit on AoA 3, both children will also have the third unit on AoA 3. The other two units, however, are allocated differently. Two equally likely pairs of offspring are possible: a new pair ([1, 1, 3] and [2, 2, 3]) and a pair that looks just like the original parents. Note that a trivial mating result is always possible.

Mutating an allocation involves the possible shifting of a unit from one AoA to another. Mutation is important because it helps the GA discover a wide range of potential allocations. In our model, each parent allocation has a 10 percent chance of moving each unit to a different AoA before mating to produce the children. Tuning this parameter is important because its value affects how rapidly the GA explores the space and how stable it is from one generation to the next. If the mutation rate is too low, the space of possible allocations may never be fully explored, whereas if the rate is too high, the GA may continually bump allocations out of their niches and may never settle down to a stable solution. We arrived at 10 percent through trial and error rather than analytical methods, but it appears to be a reasonable compromise between stability and diversity.

Defining a Field of Red Allocations

We now address Blue's knowledge of Red activity in greater detail. By activity, we mean Red's allocation of his own units to the AoAs. We measure Blue's knowledge of Red's activity on a per-unit basis by the probability that Blue knows to which AoA a Red unit has been assigned. For example, suppose that Red has three units, and Blue's activity knowledge is represented by the vector [0.0, 0.7, 0.3]; this vector specifies that Blue has no knowledge of Red unit 1's activity, that there is a 70 percent chance Blue knows the true assignment of Red unit 2, and that there is a 30 percent chance that Blue has learned the true assignment of Red unit 3. (We do not consider the possibility that Blue's beliefs are mistaken.) Because this activity vector represents additional knowledge of the scenario, it must be specified externally. More specifically,

$$\mathbf{a} = \left(a_1, a_2, a_3 \ldots a_n \right),$$

where a_i, the *activity* level for unit i, is the probability that Blue knows the AoA of Red unit i.

Based on Blue's knowledge of Red activity, we generate a field of potential Red allocations. For each Red allocation, we compare a random draw to the activity knowledge for each unit to determine if we know to which AoA the unit has been assigned. If we do not know the unit's assignment, we randomly assign it to an AoA. If we do know the unit's assignment, then we assign it to the AoA specified in the ground truth (which must be defined as part of the scenario). We repeat this process multiple times to create a field of potential Red allocations.

Fitness of a Blue Allocation

Now that we have described the generation of a field of potential Red allocations, we must address how to evaluate the fitness of a Blue allocation, which could face any of the Red allocations described in this field. The final fitness of a Blue allocation is a weighted average of basic route fitnesses over routes, strategies, and potential adversary allocations. We defined basic route fitness in Phase One as the fitness of a route for a single Blue force, given that a fixed number of Red units were assigned to oppose it in accordance with a specific strategy (intercept, mass effects, escape, etc.). Now, we must contemplate a varying number of Red units across a set of different Red strategies, and we must do this for all AoAs.

To find the fitness of a Blue allocation given a specific Red allocation, we first consider the averaging that occurs over strategies and then consider the averaging over AoAs. Recall that in Phase One, we defined the *strategic fitness* of an AoA as a weighted average of basic route fitnesses, weighted by the likelihood that Red would follow each of the possible strategies. We now define the *specific fitness* of a Blue allocation as the average of *strategic fitnesses* over all AoAs. Therefore, if m is the number of populated AoAs, then

$$F = \frac{1}{m} \sum_i \bar{f}_i.$$

It is important to note, however, that we average only over the AoAs to which Blue assigns units, excluding from the average any "empty" AoAs. We return to this decision later.

Finally, we compute the *final fitness* of a Blue allocation as the average of the *specific fitnesses*, averaged over the entire field of Red allocations. Other functions are possible. For example, one can use minimax (least regret), maximax, and others to represent risk-aversion or risk-taking on the part of the Blue commander. Blue's mission should influence this function selection for computing the fitness of a Blue allocation. The final fitness is

$$F_{allocation} = \frac{1}{n} \sum_{j=1}^{n} F_j.$$

To summarize, if we define $\bar{f}_{i,j}$ as the strategic fitness of a Blue allocation for AoA i and Red allocation j, then the final fitness is the average of the strategic fitnesses over all m nonempty AoAs and all n Red allocations:

$$F_{allocation} = \frac{1}{mn} \sum_{i=1}^{m} \sum_{j=1}^{n} \bar{f}_{i,j}.$$

The decision to exclude "empty" AoAs (AoAs to which Blue does not assign any units) from the calculation is not trivial. Consider the case in which Red assigns no units to a given

AoA. Blue will always obtain the maximum contribution to fitness from such an unguarded AoA (i.e., a fitness of 1) as long as at least one Blue unit is assigned to cover it. At first glance, then, there is no incentive for Blue to put additional units on a free AoA, even though in reality Blue would prefer to place all his units on an AoA that Red does not cover. If we do not exclude empty AoAs from the average, the fitness of the Blue allocation is reduced if Blue puts all his units on the unguarded AoA (because fitnesses of 0 are averaged in). Excluding these empty AoAs allows Blue to achieve maximum fitness by placing all his units on the unguarded AoA, thereby avoiding engagement entirely. Note that if there are two or more unguarded AoAs, the maximum fitness still averages to 1, regardless of how Blue distributes his units among the unguarded AoAs.

Summary

The Phase Two GA discovers Blue allocations by initializing a population of them and then iterating through the processes of niching, mating, mutation, and fitness evaluation. In the case of n AoAs, the Blue allocation is defined by an n-dimensional vector whose components specify the strength of all Blue units on each AoA. The niching algorithm, which is conceptually identical to the one used in Phase One, uses the Mahalanobis distance metric to determine whether two allocations (which are two points in n-dimensional space) are similar enough to mate. In Phase Two, unlike Phase One, mutation actually occurs before mating. In the mutation process, each parent allocation has a 10 percent chance of moving each unit to a different AoA before the children are produced. Mating follows according to two simple rules: If a given unit is assigned to parents with two different AoAs, it will be randomly assigned to one of those AoAs in one offspring and to the other AoA in the other offspring. If a given unit is assigned to parents with the same AoA, it will be assigned to that same AoA in both offspring. Once these allocations have been niched, mutated, and mated, we determine their fitnesses to assess which will survive into the next generation. To determine the fitness of a Blue allocation, we first define a field of potential Red allocations based on Blue's activity knowledge. The activity knowledge is defined by the probability Blue knows the AoA assignment of each Red unit. Given this knowledge, we can determine a basic route fitness for each strategy, AoA, and Red allocation. The final fitness of a Blue allocation is then the weighted average of the basic route fitnesses over routes, strategies, and potential adversary allocations.

Figure 4.2 summarizes how the final fitness of a Blue allocation can be traced through the model. The arrows trace the final fitness, in a top-down fashion, back to the Blue intelligence parameters, i.e., knowledge of Red's capability, location, intent, and activity. The basic route fitness stems from the total effect of Red on the Blue route, which derives from the location uncertainty, the effect function, and the Red behavior model. The parametric inputs to the model are the knowledge of Red activity, intent, capability, and location uncertainty; the functional inputs are the effect and uncertainty functions, as well as the entire Red behavior model. All of these inputs are highlighted by an orange box. Yellow boxes highlight the mathematical expressions that explain how final fitness depends on all these inputs.

We can also trace final fitness back to the quantitative parameters of the Blue mission, i.e., total forces allocated, willingness to engage, and starting and ending points. In addition, we can trace final fitness back to Blue's intelligence about Red, as parameterized by intent, activity, capability, location, and adaptability. The exact manner in which these factors enter the model

Figure 4.2
Tracing Final Fitness of Allocation Back Through the Model

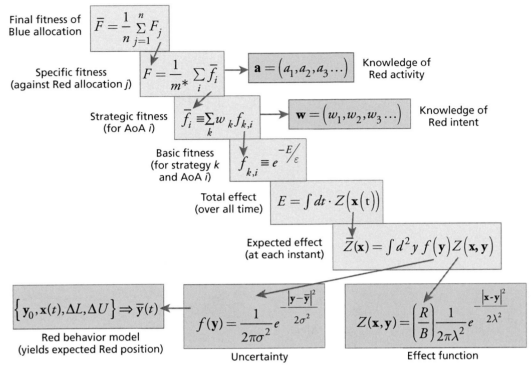

Final fitness of Blue allocation
$$\bar{F} = \frac{1}{n} \sum_{j=1}^{n} F_j$$

Specific fitness (against Red allocation j)
$$F = \frac{1}{m^*} \sum_i \bar{f}_i$$

$$\mathbf{a} = \left(a_1, a_2, a_3 \dots\right)$$ — Knowledge of Red activity

Strategic fitness (for AoA i)
$$\bar{f}_i \equiv \sum_k w_k f_{k,i}$$

$$\mathbf{w} = \left(w_1, w_2, w_3 \dots\right)$$ — Knowledge of Red intent

Basic fitness (for strategy k and AoA i)
$$f_{k,i} \equiv e^{-E/\varepsilon}$$

Total effect (over all time)
$$E = \int dt \cdot Z\left(\mathbf{x}(t)\right)$$

Expected effect (at each instant)
$$\bar{Z}(\mathbf{x}) = \int d^2 y \, f\left(\mathbf{y}\right) Z\left(\mathbf{x}, \mathbf{y}\right)$$

$$\left\{ \mathbf{y}_0, \mathbf{x}(t), \Delta L, \Delta U \right\} \Rightarrow \bar{\mathbf{y}}(t)$$
Red behavior model (yields expected Red position)

$$f(\mathbf{y}) = \frac{1}{2\pi\sigma^2} e^{-\frac{|\mathbf{y}-\bar{\mathbf{y}}|^2}{2\sigma^2}}$$
Uncertainty

$$Z(\mathbf{x}, \mathbf{y}) = \left(\frac{R}{B}\right) \frac{1}{2\pi\lambda^2} e^{-\frac{|\mathbf{x}-\mathbf{y}|^2}{2\lambda^2}}$$
Effect function

has been detailed throughout Chapters Two, Three, and Four. Figure 4.3 summarizes how the final fitness of a Blue allocation depends on Blue intelligence and mission.

The result of the model's two phases is the discovery of several feasible routes and associated allocations of Blue forces to cover those routes. In the process of computing the routes and allocations, we have incorporated Blue's intelligence about Red intent, activity, capability, location, and adaptability; we have also incorporated Blue mission parameters.

Figure 4.3
Final Fitness of Blue Algorithm Depends on Blue Intelligence and Mission

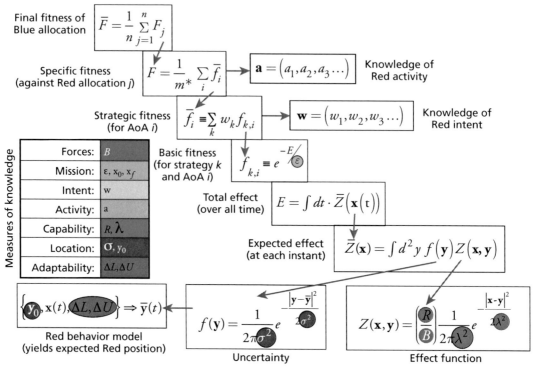

CHAPTER FIVE

Modeling Terrain

Until now, our calculations have assumed that the effect function depends only on the distance between Red and Blue units. As previously noted, this implicitly assumes a lack of interesting terrain, an assumption rarely borne out by reality. Next, we incorporate the influence of terrain on the combat engagement. There are several places in the model where terrain can be incorporated naturally. We introduce three specific measures of terrain into the analysis: impassibility, inhospitableness, and shadowing. "Impassibility" measures the degree to which Blue finds certain terrain difficult to cross. "Inhospitableness" measures the degree to which Blue believes that terrain will affect Red's choice of location. "Shadowing" models the influence of terrain on Red's effect on Blue.[1] We now describe each of these parameters in more detail and show their effects on fitness calculations in the GA.

Impassibility

Impassibility measures the degree to which Blue finds certain terrain difficult to cross. For example, dense forest, steep ground, and even an area deemed to be risky (e.g., covered in land mines) could result in or contribute to impassibility. Note that although it seems natural to do so, our model does not actually impede Blue progress in an "impassible" area. This is because we are not explicitly modeling Blue's changes in speed. However, this approximation should be lifted in the future, since speed is a crucial aspect of maneuver.

Instead, we define impassibility for every point and integrate it along an entire Blue route to determine the overall difficulty of traversing the route. We define a function $T(x)$ to represent the difficulty of passage through location x. This function takes on values between zero and 1. A value of zero occurs at locations x that Blue may pass through freely. A value of one occurs at locations y that Blue cannot pass through. We then define the total impassability of a route as the integral of T over all points x in the route. Thus,

$$T = \int dt \cdot T\big(\mathbf{x}(t)\big).$$

Like total effect, total impassability influences route fitness. We define the relation between impassability and route fitness by a decaying negative exponential function whose exponent is equal to total impassability weighted by the tuning parameter $(1/\tau)$. The parameter τ can be

---

[1] Terrain could also influence (1) Blue's ability to deduce Red's activity and intent and (2) Red's adaptability. These effects are not incorporated into our current model.

used to tune Blue's willingness to pass through undesirable terrain. (Note that nonzero impassibility will implicitly penalize longer routes.) We modified the definition of basic route fitness to incorporate terrain impassability:

$$f \equiv e^{-\left(E/_\varepsilon + T/_\tau\right)}.$$

The parameters ε and τ now represent the relative importance of Red exposure compared to terrain exposure. Specifically, the parameter ε represents Blue's willingness to be exposed to Red effect; τ represents Blue's willingness to travel over difficult terrain. These two parameters can be adjusted to reflect the relative importance that Blue assigns to each. To be fit, a route must be easy to traverse *and* must limit Blue's exposure to Red's expected effect.

Inhospitableness

Inhospitableness represents Blue's perception of how terrain will affect Red's choice of location. Inhospitableness is not the same as impassibility, but the two are clearly related in many cases where certain aspects of terrain (e.g., steepness) affect both Red and Blue similarly.

We define a function $L(\mathbf{y})$ to represent Blue's perception of how unlikely it is to find Red at location $\mathbf{y}$ due to terrain factors. This function takes on values between zero and one. A value of zero occurs at locations $\mathbf{y}$ whose terrain, Blue believes, will not deter Red's ability or willingness to occupy the location. A value of one occurs at locations whose terrain, Blue is certain, will prevent Red from occupying the location. For example, an exceptionally steep terrain feature may preclude any heavy units from occupying territory, and this information can be incorporated with a value of close to 1. We then modify the uncertainty function for Red's location, $f(\mathbf{y})$, to incorporate the inhospitableness as described by $L(\mathbf{y})$. Thus,

$$f(\mathbf{y}) = \frac{1}{2\pi\sigma^2} e^{-\frac{|\mathbf{y} - \bar{\mathbf{y}}|^2}{2\sigma^2}} \times \left(1 - L(\mathbf{y})\right).$$

While in principle we ought to renormalize the overall function $f(\mathbf{y})$ to ensure that it is still a valid density function (it must integrate to 1 over all space), that step is not crucial. An overall constant will not affect relative fitness but will enter into the equation when we compare the relative importance of Red effect and terrain impassability. In that case, a lowered total effect due to the modification just described must be compensated for by lowering or raising the respective parameters.

Ideally, we would like to generate both impassibility and inhospitableness from a single function of terrain. This is sensible for a simple facet of terrain, such as a mountain, and we do this in the model as implemented.

Shadowing

The third part of our representation of terrain is the effect of terrain on Red's *effect*. We refer to this phenomenon as "shadowing" and note that it is different from the effect of terrain on Red itself. Hence, it must be modeled separately. In some cases, shadowing may be attributed solely to Blue's position. For example, a jungle may provide cover for Blue. In other cases, a mountain may provide cover depending on the relative positions of Red, Blue, and the ridgeline.[2]

We define a function $M(\mathbf{x},\mathbf{y})$ that represents the influence of terrain on Red's ability to affect location $\mathbf{x}$ from location $\mathbf{y}$. This function takes on values between zero and one. A value of zero implies that Red cannot affect location $\mathbf{x}$ from location $\mathbf{y}$; a value of one implies that terrain has no effect on Red's ability to affect location $\mathbf{x}$ from location $\mathbf{y}$. We then modify the effect function by multiplying it by $M(\mathbf{x},\mathbf{y})$. Thus, the modified effect function is

$$Z(\mathbf{x},\mathbf{y}) = \left(\frac{R}{B}\right)\frac{1}{2\pi(\lambda^2)}e^{-\frac{|\mathbf{x}-\mathbf{y}|^2}{2\lambda^2}} \cdot M(\mathbf{x},\mathbf{y}).$$

A Terrain Example

We consider a simple terrain example: a mountain feature that affects Red and Blue in a similar manner. For simplicity, we represent the mountain as a modified Gaussian,

$$T(x) = e^{-\frac{(\mathbf{x}-\mathbf{a})\cdot M \cdot (\mathbf{x}-\mathbf{a})}{2\eta^2}},$$

where

$$\mathbf{M} = \begin{pmatrix} \cos(\theta) & \sin(\theta) \\ -\sin(\theta) & \cos(\theta) \end{pmatrix}\begin{pmatrix} p & 0 \\ 0 & 1 \end{pmatrix}\begin{pmatrix} \cos(\theta) & -\sin(\theta) \\ \sin(\theta) & \cos(\theta) \end{pmatrix}.$$

In this case, $\mathbf{a}$ defines the center of the mountain, and η is its width. $\mathbf{M}$ stretches the mountain into a ridge and rotates it as needed, p is the mountain's elongation (i.e., aspect ratio), and θ is the direction of the ridgeline. An example of such a terrain feature is shown in Figure 5.1. Multiple Gaussians can be added together to create a more complicated landscape.

---

[2] As an effect on an effect, shadowing is something of a second-order consideration. Accordingly, our current model does not currently implement the shadowing function.

Figure 5.1
An Example of a Mountain Ridge Terrain Feature

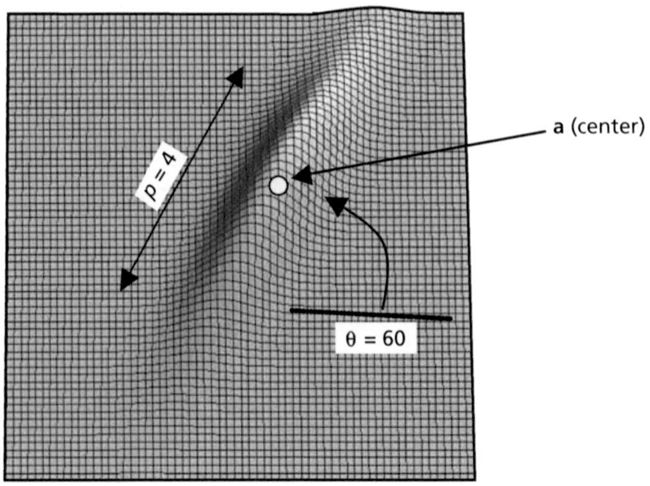

RAND *TR423-5.1*

Summary

Table 5.1 explains how terrain is introduced into the model. In particular, we show how impassability, inhospitableness, and shadowing ultimately affect the basic fitness of a route. In the next chapter, we give numerous examples of the GA outputs, including ones that demonstrate the effect of terrain on AoA selection. For simplicity, all terrain examples employ the same function to represent impassability, $T(\mathbf{x})$, and inhospitableness, $L(\mathbf{y})$, using the modified Gaussian

$$T\left(x\right)=e^{-\dfrac{\left(x-a\right)\cdot M\cdot\left(x-a\right)}{2\eta^2}}.$$

Table 5.1
Summary of Model Extensions to Incorporate Terrain Effects

Terrain Parameter	Effect on Model	Comments
Impassibility—$T(\mathbf{x})$	$$f \equiv e^{-\left(E/_\varepsilon + T/_\tau\right)},$$ where $$T = \int dt \cdot T\big(\mathbf{x}(t)\big)$$	Route fitness now depends both on exposure to Red's effect and traversal of undesirable terrain
Inhospitableness—$L(\mathbf{y})$	$$f(\mathbf{y}) = \frac{1}{2\pi\sigma^2} e^{-\frac{|\mathbf{y}-\overline{\mathbf{y}}|^2}{2\sigma^2}} \times \big(1 - L(\mathbf{y})\big)$$	Red's location uncertainty now depends on Blue's perception of how inhospitable the terrain is
Shadowing—$M(\mathbf{x},\mathbf{y})$	$$Z(\mathbf{x},\mathbf{y}) = \left(\frac{R}{B}\right)\frac{1}{2\pi(\lambda^2)} e^{-\frac{|\mathbf{x}-\mathbf{y}|^2}{2\lambda^2}} \cdot M(\mathbf{x},\mathbf{y})$$	Effect function now depends not only on Red's strength and range but also on the shadowing effects of terrain (e.g., those caused by mountains or jungles)

Proof-of-Principle Examples

The examples provided in this chapter assume that Blue's mission is to move from one place to another while avoiding Red units (enemies). Our examples sometimes use dots to depict unit locations and sometimes use topographic landscapes for ease of visualization. However, these examples could just as easily have been drawn with a more military flavor, as demonstrated in Figure 6.1. In this figure, Blue has been instructed to move to point alpha while avoiding Red units, represented by red diamonds, along the way. In the examples that follow, we consider simple bypass scenarios and show the outputs of the GA, which discovers potential Blue AoAs and the effects that Blue is likely to encounter.

Figure 6.1
An Example of a Blue Bypass Mission

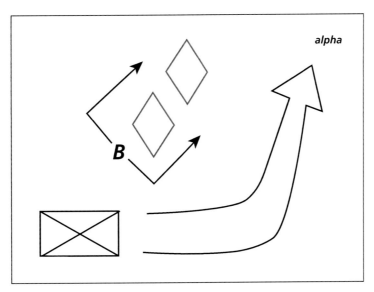

A Simple Scenario

In the simple scenario depicted in Figure 6.2, Red remains stationary at the midpoint between Blue's starting position (A) and Blue's destination (B). Blue has multiple options to consider when planning its route. Ultimately, it is the distance between Blue and Red, compared to the characteristic distance associated with Red's expected effect (the standard deviation of the Gaussian that represents it), that determines Blue's best option. As previously shown, the widths of the uncertainty function and the Red effect function combine in the expected effect expression; the final standard deviation of expected effect is the square root of the sum of variances, which we refer to as ρ.

There are three cases to consider. If both the start and end points (A and B) lie within ρ of Red's expected position, then Blue's knowledge of Red's position is so uncertain that Blue's best option (i.e., the one that minimizes total expected effect) is simply to take the shortest route possible to reach his destination. On the other hand, if A and B are farther than ρ from Red's expected position, then Blue knows enough about Red's location that he benefits from maneuvering around Red. In that case, Blue should take the longest route to reach his destination. Finally, if the distances in question are all similar to the characteristic distance, each solution is equally viable.[1]

This simple example illustrates a primary tradeoff: uncertainty about expected effect along an AoA versus the directness of a route. If Blue's knowledge of Red's whereabouts is good, then Blue benefits by taking time to maneuver around Red. On the other hand, if Blue's knowledge of Red's location is poor, then time is more critical, and Blue's best option is perhaps to make a beeline for his destination.

Although both the location uncertainty width, σ, and the capability width, λ, play a role in determining the width of the expected effect map, we consider conceptually in the results that follow only the effects of varying σ. Analogous comments could be made about the effect of variations in λ because it combines the two variances that determine the expected effect.

Figure 6.2
A Simple Scenario

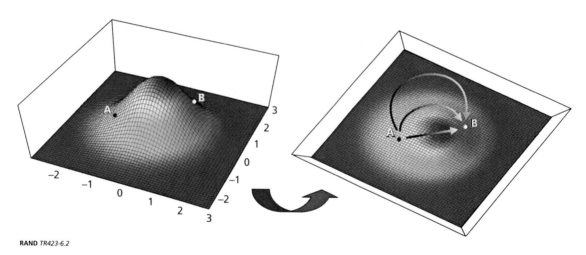

RAND TR423-6.2

---

[1]   The articulation of the three possible options is the direct result of a simple scenario in which the mathematics are easily calculated. Thus we were able to verify that the model as instantiated was working correctly.

We parameterize our results by the relative uncertainty in Red's position, where relative uncertainty is defined as the ratio of the uncertainty in Red's position, σ, to the average distance, d, that Blue would be from Red if he were to take the shortest path from A to B.

The Case of High Uncertainty

Here we examine the results of the GA from the simple scenario described above, this time assuming high relative uncertainty about Red's position. In this case, σ/d equals 1.4. The upper plots in Figure 6.3 depict the top ten fittest paths, and the lower plots depict the full set of centroids of all paths in a generation for generations zero, ten, and 20. After only ten generations, the top ten fittest paths have all converged on the straight-line, shortest-distance path between A and B, as expected. If the location uncertainty width, σ, is high relative to the average distance, d, that Blue would be from Red (on the straight-line path), then the expected effect map will be relatively flat compared to the one generated using a smaller σ. Hence, Blue does not benefit from maneuvering around Red, instead preferring the shortest path.

In the centroid plots, A and B are the start and end points of all Blue paths, and the red dot represents Red's position (which, in this simple case, is stationary). The Blue dots represent the centroids of the entire population of Blue paths, and the purple star represents the centroid of the centroids. The larger blue ellipse depicts the niche radius, used to determine whether two paths are sufficiently similar to be suitable mates. As the GA converges, the set of centroids starts to collapse to a single point. Note that in generation 20, there are multiple blue dots piled on top of one another near the centroid of the centroids.

Figure 6.3
Path Evolution—High Uncertainty

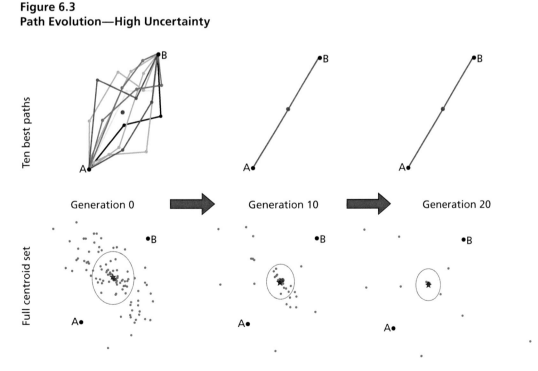

NOTE: At high uncertainty, σ/d=1.4, the paths quickly converge on one solution. I.e., Blue goes straight through.

RAND TR423-6.3

The Case of Moderate Uncertainty

Figure 6.4 shows the results of the GA, this time assuming moderate relative uncertainty about Red's position. In this case, σ equals d, and all three solutions are viable options for Blue. At generation 20, the top ten fittest paths include both options to maneuver around Red on either side as well as the shortest path between A and B. The corresponding sets of centroids over the generations shows the emergence of three distinct niches (i.e., groups of centroids), as expected.

Figure 6.4
Path Evolution—Moderate Uncertainty

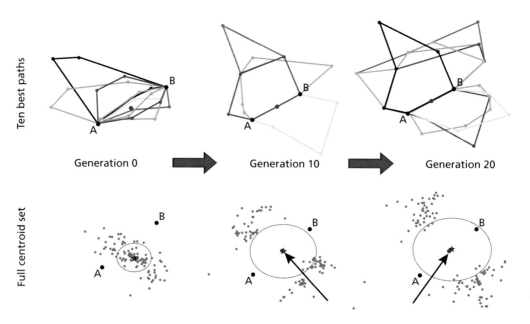

NOTES: At moderate uncertainty, σ/d=1, all three solutions emerge. Multiple paths are coextensive. Several Blue path centroids are piled up in the middle and difficult to see in the plots.

RAND TR423-6.4

The Case of Low Uncertainty

Figure 6.5 shows the results of the GA, this time assuming low relative uncertainty about Red's position. The paths rapidly diverge, generating two distinct solutions that show Blue's options for maneuvering around Red. This result is not surprising, since Blue has good knowledge of Red's whereabouts and hence benefits by maneuvering around him.

Figure 6.5
Path Evolution—Low Uncertainty

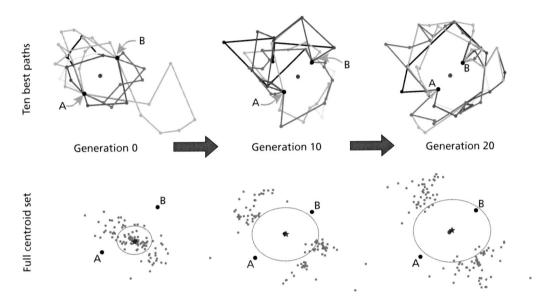

NOTE: At low uncertainty, $\sigma/d = 0.5$, the paths rapidly diverge into two solutions. I.e., Blue goes around Red.

RAND *TR423-6.5*

Final AoAs

Here we examine the AoAs discovered by the GA after 20 generations. We also study the effect profiles of these "final AoAs" for the cases of high, medium, and low relative uncertainty about Red's location. We stop the GA after 20 generations because of the relative stability of the population's average fitness of routes in the medium and high uncertainty cases. As noted below, fitness does not stabilize after 20 generations in the case of low uncertainty, but the fit AoAs are clearly those that maneuver around Red.

Figure 6.6 shows the results of the high relative uncertainty case, in which the fittest path, the green path, is the shortest one to the destination. The effect profiles confirm that this path does indeed have the smallest total effect (i.e., area under the curve). However, since this model is a planning tool, it is interesting to look at other potential AoAs and the effects that Blue would expect to encounter if he chose those paths instead.

Figure 6.6
Final AoAs—High Uncertainty

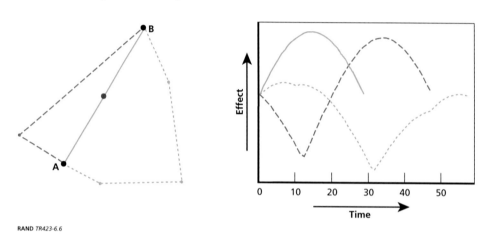

RAND TR423-6.6

Figure 6.7 shows the results of the moderate relative uncertainty case, in which all three AoAs are viable options for Blue. Although the fitnesses of these AoAs are comparable, their effect profiles are considerably different from one another. Which path is preferable depends on

Figure 6.7
Final AoAs—Moderate Uncertainty

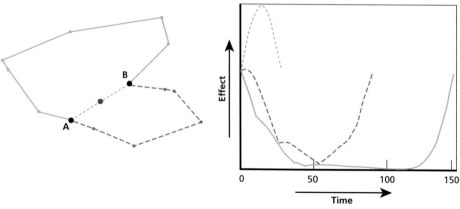

RAND TR423-6.7

the interpretation of the effect profile. In our model, we consider only the area under the curve, but many other metrics could also be useful.

Figure 6.8 shows the results of the low relative uncertainty case, in which the GA finds two distinct AoAs of nearly equal fitness. Blue has the option of maneuvering around Red to either side. Due to the symmetry of the scenario, the effect profiles are similar even though the AoAs are distinct.

Figure 6.8
Final AoAs—Low Uncertainty

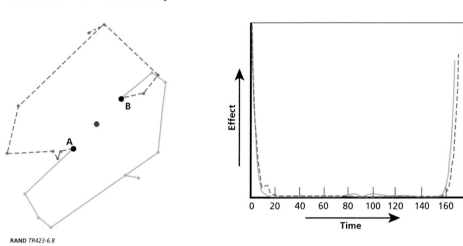

RAND *TR423-6.8*

Fitness Evolution

Figure 6.9 depicts the fitness evolution for the simple scenario run under the cases of high, medium, and low relative uncertainty. The high-uncertainty case on the left corresponds to poor knowledge about Red. In this case, the convergence to the degenerate solution is rapid. Because Blue does not have significant information regarding Red, he is unable to make any plan other than the most obvious decision to go straight to his destination.

Figure 6.9
Fitness Evolution

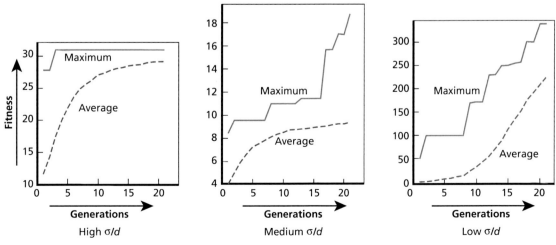

RAND *TR423-6.9*

In the case of low uncertainty, Blue is so certain of Red's position that he can choose a clear path around Red. Because our current scenario does not directly penalize Blue for expending more time, taking a path far away from Red is desirable. The solution diverges to infinity in this very simple case.

In case of medium uncertainty, a combination of the behavior exhibited in the low- and high-uncertainty cases emerges. The large jump at generation 15 represents the "discovery" of a new niche and hence an improvement in fitness.

Terrain Effect on AoA Selection

Our next example demonstrates how terrain influences the selection of Blue AoAs. We consider a simple scenario with two Red units. On the left side of Figure 6.10, three AoAs are diverted south of the two Red units. On flat land, Blue's best choice is to go straight to the goal, simply avoiding Red along the way.

The mountain shown on the right side of Figure 6.10 is not technically impossible, but it increasingly penalizes both Red and Blue forces as they approach the ridge's center. Red is not expected to travel over the mountain, so the effect function is lowered there; ordinarily, this would make the mountain more attractive to Blue. However, Blue's route fitness is penalized by the difficult terrain as well. Faced with this mountain, Blue may skirt its base to the right or go well out of the way to the left to avoid Red.

In this instance, the Red units are moving slower than Blue and have moderate adaptability. We have chosen this simple scenario, with only two Red units, to illustrate the terrain effects more clearly. In more-complicated scenarios, it can be difficult for the eye to discern the effects.

Figure 6.10
Terrain Influence on AoAs

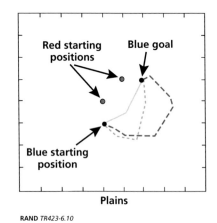
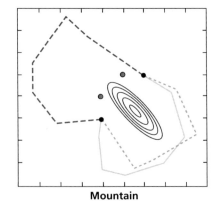

Plains   Mountain

Figure 6.11 shows the anticipated Red response to each of the three Blue AoAs. In this scenario, the Red strategy is for each unit to attempt to intercept Blue individually. Red is slightly slower than Blue and has moderate to low adaptability. Note that Red is willing to traverse some of the lower parts of the mountain, but only when he can move significantly closer to Blue. For example, in AoAs 1 and 2, Red is willing to cross at higher elevations of the mountain near the start of his path when he is closer to Blue. However, in the middle of Red's path, he is farther behind Blue and hence has less to gain by traversing the mountain. Red chooses his direction by trying to maximize his expected effect, but he encounters a tradeoff: quickly gaining proximity to Blue by going through worse terrain versus slowly catching up to Blue but going through better terrain.

Figure 6.11
Red Pursues Blue Around the Mountain

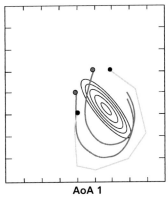
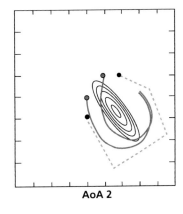
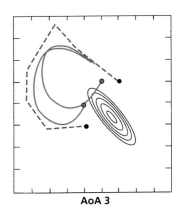

AoA 1   AoA 2   AoA 3

Effect of Red Behavior Model on AoA Selection

Our next set of examples demonstrates how Red's behavior model, specifically Blue's perception of Red's adaptability and intelligence, influences both Blue's AoA selection and Red's motion. Here we consider a new scenario in which two Red units are trying to coordinate their efforts to mass effects on Blue.

Effect of Red's Adaptability

In the case of a barely adaptive Red, Blue tricks Red into following him around; accordingly, Red is unable to intercept Blue, as seen in Figure 6.12. Note that while the Blue starting position and mission are the same as in the example used to illustrate the effects of terrain, the Red units start in a different position. This particular engagement takes place on flat terrain. We show the results for Red units that can travel at 60 percent of Blue's speed.

In this example, updates occur at every four time steps, giving Red a fairly low ability to adapt. To put that number in perspective, if Blue heads straight for his destination, he will reach it within about 30 time steps. Most paths are significantly longer, as shown. Red's intelligence is neither gaining ground nor failing to keep up, because Red receives exactly four new time steps of information every four time steps. In terms of earlier notation, the parameter D equals 1.

Figure 6.12
Blue Evades Red When Red Is Less Adaptive

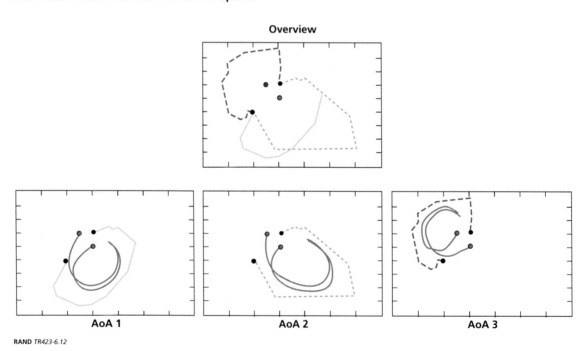

Figure 6.13 shows the results for a more adaptive Red. Here, Red is receiving updates about Blue's path every two time steps, twice as frequently as during the previous case. Every two time steps, Red acquires two time steps' more worth of knowledge regarding Blue's future path. Note that Red stops when he intercepts Blue's path, but will start up again and follow if Blue runs past. In this case, the most fit path is AoA 1, which goes straight to the destination. Because Blue realizes that he cannot avoid Red, Blue's best course of action is the shortest path.

Remember, however, that the adaptability parameter describes Blue's perception of Red's behavior. If Blue credits Red with being smarter than he really is, Blue's options diminish. This is precisely what is occurring in Figure 6.13 compared to Figure 6.12.

Figure 6.13
Blue Cannot Escape a More Adaptive Red

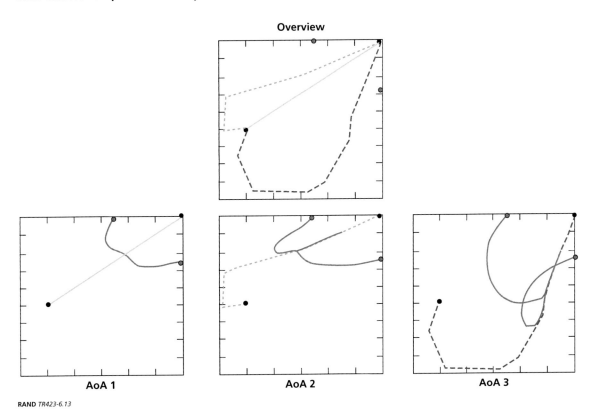

Figure 6.14 shows the effect profiles of the three AoAs for the less adaptive (left) and more adaptive (right) cases. In the more adaptive case, Blue is exposed to high levels of Red effects, but this exposure occurs over a shorter period of time, since Red is able to intercept Blue along a more direct route. In the less adaptive case, Blue takes a much longer route while avoiding Red; therefore, Blue is exposed to low levels of Red effect, but this exposure occurs over a longer period of time.

In the less adaptive case, the effect profile shows that Red gets close to Blue at the beginning and end of his journey, but Blue manages to avoid Red during the middle; in the more adaptive case, Red always intercepts.

Figure 6.14
Effect Profiles Reflect the Results of Red Adaptability

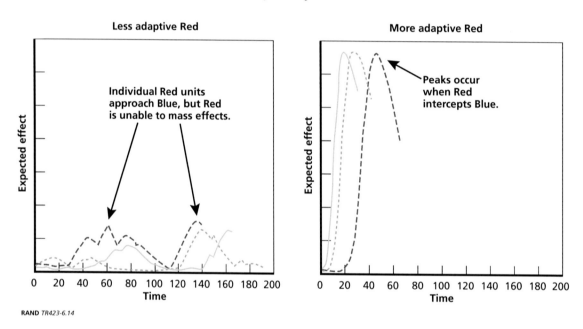

Effect of Red's Intelligence

Here we demonstrate the impact of Red's intelligence on Red's motion and hence ultimately on AoA selection. We used the same scenario described in the adaptability examples but kept the time between updates fixed at every two time steps, varying instead the look-ahead times. Figure 6.15 compares the AoAs and corresponding Red responses as Red intelligence increases. As Red gains more intelligence, Blue realizes that he cannot evade Red and instead heads directly for his destination. Red is more likely to intercept Blue as Red gains more intelligence. Red also intercepts Blue sooner, with fewer changes in direction.

Figure 6.15
Increasing Red Intelligence Diminishes Blue Options

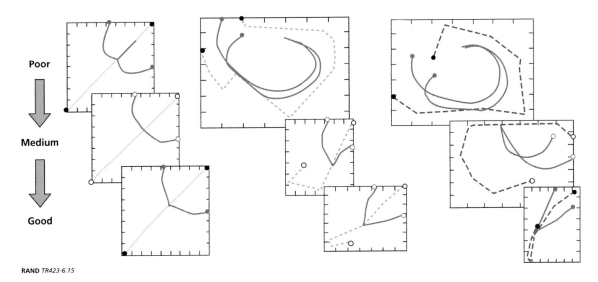

RAND *TR423-6.15*

Red Activity Knowledge Affects Blue Allocations

When Blue has no knowledge, the field of allocations he faces is quite varied, and Blue will tend to allocate forces according to fitness. In the example shown in Figure 6.16, AoA 1 is the most fit and AoA 3 is the least fit. Hence, Blue allocates a majority of his forces to AoA 1, and fewer and fewer forces to AoAs 2 and 3. If Blue has some knowledge of Red activity (the chart in the middle), then Blue will try to dominate Red where he expects Red to go and will allocate forces where he expects Red not to be. We see precisely this effect in the case when Blue knows where only half of Red's units are. With full knowledge (the chart on the right), Blue can successfully plan to avoid the main Red force.

Figure 6.16
Improvement in Blue Allocations with Red Activity Knowledge

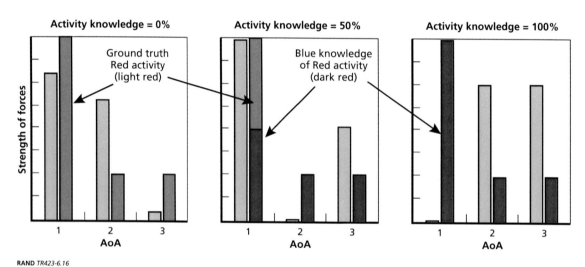

Summary

To validate the model, we first considered a simple scenario in which Red was stationary and located at the midpoint between Blue's start and end points. Blue's mission was to reach his destination while minimizing his exposure to Red. Routes discovered by the model show that the chosen path depends on intelligence about Red's location and capability. Specifically, if Blue is relatively certain about Red's location and capability, then Blue benefits from maneuvering around Red. However, if Blue is very uncertain about Red's location or capability or both, Blue does not benefit from maneuvering around the enemy, but rather should take the most direct route to his destination. Over various cases, the model clearly demonstrates the value of intelligence in the planning process.

The model also demonstrates the influence of terrain on AoA selection. We modeled a simple mountain feature that affected both Blue and Red forces. Because Red was less likely to travel over the mountainous area, his effect on Blue was diminished in this region. Hence, we expected this region to be more desirable to Blue. However, the desirability of Blue routes was also penalized by the difficult terrain. Ultimately, the model discovered Blue AoAs that avoided the mountainous region as much as possible while minimizing Blue exposure to Red.

We also considered various other cases to demonstrate the effect of enemy intelligence and adaptability on AoA selection. Enemy adaptability is the rate at which Red receives updates about Blue's route. Enemy intelligence is the amount of information Red receives at each update. Unsurprisingly, Blue is more likely to evade a less adaptive Red than a more adaptive Red. Also, more-intelligent Red forces diminished Blue's options.

We also demonstrated the effect of Blue knowledge of Red activity on Blue force allocation to AoAs. Activity knowledge is the likelihood that Blue knows to which AoA each Red unit has been assigned. We found that with partial activity knowledge, Blue tries to dominate Red where Blue expects Red to be; Blue will also allocate forces where he expects Red not to be. With perfect activity knowledge, Blue can completely avoid the AoA where the dominant Red force is located.

CHAPTER SEVEN

Conclusions and Future Extensions

This report describes a combat planning model that uses GAs to discover Blue AoAs and associated force allocation schemes. The model uses both Blue's intelligence about Red and Blue's intelligence about the terrain to perform route planning and force allocation. Blue's intelligence about Red consists of knowledge of Red's location, capability, activity, and intent. Blue's knowledge of the terrain is parameterized by his knowledge of the impassibility, inhospitableness, and shadowing effects of his environment. Knowledge of Blue's mission, specifically his start and end points and his willingness to engage, also play a role in the planning process.

We also developed a Red behavior model to enable planning against a sophisticated enemy. This model includes parameters that allow us to quantify Blue's perception of Red's intelligence and adaptability. These parameters affect Red's motion and hence Blue's ultimate AoA selection. The sophisticated look-ahead representation used in Red's behavior model allows the model to discover interesting Blue maneuver schemes, including feints.

The landscape of possible Blue AoAs and allocations is sufficiently vast that a smart search algorithm that can evolve solutions to this problem is necessary. The GA approach proved an efficient means of searching this space. One general implementation challenge associated with the method described here lies in generating data to populate the model. Specifically, it is important to choose both the appropriate "capability" index for both Blue and Red as well as ways to translate that capability into "fitness." In our examples, we used generalized capabilities and an additive function to determine overall fitness.

Future extensions of this combat planning model are possible. For instance, we could improve the modeling of Blue options and Red's adaptability. The Red behavior model could include more-complex Red projection of Blue movement for the cases in which Red lacks knowledge of Blue's future path. We could also extend the model to allow the speed of Red and Blue units to vary during route planning. Finally, the extrapolation of various Blue missions would be a useful extension.

Simultaneous consideration of AoAs would allow us to consider synergistic effects between Red and Blue units. Where AoAs converge or cross, units pursuing one AoA should be able to affect units on the other AoA. This extension would give further insights into AoA selection and allow synchronization in planning schemes to be explored.

References

Allen, Patrick D., *Situational Force Scoring: Accounting for Combined Arms Effects in Aggregate Combat Models*, Santa Monica, Calif.: The RAND Corporation, N-3423-NA, 1992. As of April 17, 2008:
http://www.rand.org/pubs/notes/N3423/

Bass, Tim, "Intrusion Detection Systems and Multisensor Data Fusion," *Communications of ACM*, Vol. 43, No. 4, 2000, pp. 99–105.

Bellman, Richard Ernest, Robert E. Kalaba, and L. A. Zader, *Abstraction and Pattern Classification*, Santa Monica, Calif.: RAND Corporation, RM-4307-PR, 1964. As of April 17, 2008:
http://www.rand.org/pubs/research_memoranda/RM4307/

Davis, Paul K., *Aggregation, Disaggregation, and the 3:1 Rules in Ground Combat*, Santa Monica, Calif.: The RAND Corporation, MR-638-AF/A/OSD, 1995. As of April 17, 2008:
http://www.rand.org/pubs/monograph_reports/MR638/

Filippidis, A., L. C. Jain, and N. Martin, "Multisensor Data Fusion for Surface Land-Mine Detection," *IEEE Transactions on Systems, Man, and Cybernetics, Part C: Applications and Reviews*, Vol. 30, No. 1, February 2000, pp. 145–150.

Hart, P. E., N. J. Nilsson, and B. Raphael, "A Formal Basis for the Heuristic Determination of Minimum Cost Paths," *IEEE Transactions on Systems Science and Cybernetics*, Vol. 4, No. 2, July 1968, pp. 100–107.

Jaiswal, N. K., *Military Operations Research: Quantitative Decisionmaking*, Boston, Mass.: Kluwer Academic Publishers, 1997.

Kewley, R. H., and M. J. Embrechts, "Fuzzy-Genetic Decision Optimization for Positioning of Military Combat Units," *IEEE International Conference on Systems, Man, and Cybernetics*, Vol. 4, No. 11, October 1998, pp. 3,658–3,664.

Krieg, Mark L., "A Tutorial on Bayesian Belief Networks," Edinburgh, South Australia: Defence Science and Technology Organisation Electronics and Surveillance Research Laboratory, DSTO-TN-0403, December 2001. As of April 17, 2008:
http://www.dsto.defence.gov.au/publications/2424/DSTO-TN-0403.pdf

Liang, Y., F. Robichaud, B. Fugere, and K. Ackles, "Implementing a Naturalistic Command Agent Design," in *Proceedings of the 10th Conference on Computer Generated Forces and Behavioral Representation*, Norfolk, Va.: SISO, Inc., 2001, pp. 379–386.

Mitchell, Melanie, *An Introduction to Genetic Algorithms*, cloth ed., Cambridge, Mass: The MIT Press, 1996.

Moriarty, David E., "Determining Effective Military Decisive Points Through Knowledge-Rich Case-Based Reasoning," in Rasiah Loganantharaj, Gunther Palm, and Moonis Ali, eds., *Intelligent Problem Solving—Methodologies and Approaches: 13th International Conference on Industrial Engineering Applications of Artificial Intelligence and Expert Systems, IEA/AIE 2000, New Orleans, Louisiana, USA, June 19–22, 2000 (Proceedings)*, Secaucus, N.J.: Springer, 2000.

Murray, Col Charles H., ed., *Executive Decision Making*, 6th ed., Newport, R.I.: U.S. Naval War College, February 1, 2002.

Pernin, Christopher G., and Louis R. Moore, *The Weapons Mix Problem: A Math Model to Quantify the Effects of Internetting of Fires to the Future Force*, Santa Monica, Calif.: The RAND Corporation, TR-170-A, 2005. As

of April 17, 2008:
http://www.rand.org/pubs/technical_reports/TR170/

Pernin, Christopher G., Louis R. Moore, and Katherine Comanor, *The Knowledge Matrix Approach to Intelligence Fusion*, Santa Monica, Calif.: The RAND Corporation, TR-416-A, 2007. As of April 17, 2008: http://www.rand.org/pubs/technical_reports/TR416/

Qin, Qiming, Robert R. Gillies, Rongjian Lu, and Shan Chen, "An Integration of Wavelet Analysis and Neural Networks in Synthetic Aperture Radar Image Classification," in Orhan Altan, ed., *Geo-Imagery Bridging Continents: XXth ISPRS Congress Proceedings*, Volume XXXV, Part B2, 2004, p. 181 ff. As of April 17, 2008:
http://www.isprs.org/istanbul2004/comm2/comm2.html

Rogers, Steven K., John M. Colombi, Curtis E. Martin, James C. Gainey, Ken H. Fielding, Tom J. Burns, Dennis W. Ruck, Matthew Kabrisky, and Mark Oxley, "Neural Networks for Automatic Target Recognition," *Neural Networks*, Vol. 8, No. 7–8, 1995, pp. 1,153–1,184.

Starr, Colin, and Peng Shi, "An Introduction to Bayesian Belief Networks and Their Applications to Land Operations," Edinburgh, South Australia: Defence Science and Technology Organisation Systems Sciences Laboratory, DTSO-TN-0534, January 2004. As of April 17, 2008:
http://www.dsto.defence.gov.au/publications/2655/DSTO-TN-0534.pdf